Current Topics in Neuroendocrinology Volume 1

Sleep Clinical and Experimental Aspects

Editors

D. Ganten and D. Pfaff

Contributors

A. A. Borbély R. Drucker-Colin

J. V. Jones P. Sleight H. S. Smyth G. Stock

C. Valverde-R.

With 47 Figures

Springer-Verlag

Berlin Heidelberg New York 1982

Editors

Dr. Detlev Ganten, M. D., Ph. D.
Pharmakologisches Institut
Universität Heidelberg
Im Neuenheimer Feld 366
D-6900 Heidelberg/FRG

Dr. Donald Pfaff, Ph.D.
Rockefeller University
York Avenue, and 66th Street
New York, NY 10021/USA

The picture on the cover has been taken from Nieuwenhuys R., Voogd J., van Huijzen Chr.: The Human Central Nervous System. 2nd Edition. Springer-Verlag Berlin Heidelberg New York 1981

ISBN-13: 978-3-642-68335-0 e-ISBN-13: 978-3-642-68333-6
DOI.10.1007/978-3-642-68333-6

Library of Congress Cataloging in Publication Data. Main entry under title: Sleep, clinical and experimental aspects. (Current topics in neuroendocrinology; v. 1) Includes bibliographies and index. 1. Sleep–Physiological aspects. 2. Sleep disorders. I. Borbély, Alexander A. II. Ganten, D. (Detlev), 1941-. III. Pfaff, Donald W., 1939-. IV. Series [DNLM: 1. Sleep–Physiology. WI CU82Q v. 1/WL 108 S62] QP425.S674 612'.821 81-23294 AACR2

Typesetting, printing, and bookbinding by Brühlsche Universitätsdruckerei Giessen
2121/3130-543210

Preface

There is no doubt that a major problem of present day research workers, especially in the life sciences, is the plethora of publications of all kinds, abstracts, short communications, full papers in journals of varying quality, reviews and proceedings of symposia with, in addition, an unprecedented duplication of publications. Even for experts working in the field, it is almost impossible to keep an up-to-date view of all current research articles. The Western grant and career system encourages scientists to publish as much as possible.

The editors and publishers of our new series are convinced that the format of Current Topics in Neuroendocrinology leads a way out of this confusion. Each volume is conceived as a concise up-to-date textbook on one well-defined and currently exciting subject. Different from classic textbooks, however, the speed of publication compares favorably with that of many journals; this ensures an immediacy which is impossible in textbooks. On the other hand, topics to be included in this series are also sufficiently reliable, with enough work being done to treat them from several aspects. Each volume will supply four to six chapters treating such a broad topic as neuroendocrinology from several points of view, for example, anatomic, electrophysiologic, endocrine and behavioral views. Where clinical data are immediately available, they will be included. No other series treating the nervous or endocrine systems provides such a coordinated set of chapters on an interesting topic in each volume. Readers can expect authoritative reviews of subjects which are developing quickly within the often complex fields of integrative neurophysiology and experimental endocrinology. For researchers and students, this gives the chance to supplement their ability to cover the primary research literature. For lay persons and clinicians, this updates, for the most interesting topics, what they would have read in the most recent textbooks. As opposed to the frequent proceedings of symposia with a usually rather heterogenous collection of original and review papers, the reader will "know what he is getting" from the title of each volume of Current Topics in Neuroendocrinology, even without knowing much about the individual authors. It will be our objective as editors to supply the reader of this series with a concise, critical, and comprehensive collection of chapters on one specific topic in neuroendocrinology.

Volume I of the series Current Topics in Neuroendocrinology covers mechanisms of Sleep from a neurochemical, endocrine, and clinical point of view. In four chapters, in-depth discussion include of the neurotransmitters, pharmacologic interferences, endocrine, and peptidergic functions, a characterization of Sleep with regard to hemodynamics, single unit recordings and EEG activity as well as biologic rhythms and sleep characterization in patients. Each chapter is preceded by a table of contents which enables the reader to orient himself quickly as to where to look and what to expect.

Professor Günter Stock from the Department of Physiology, University of Heidelberg, has given us his expert advice in putting this first volume together and reviewing the manuscripts. We also wish to acknowledge the efficient help of Springer Verlag for the edition of this series. We hope that Current Topics in Neuroendocrinology fulfills the needs of the scientific community for concise and comprehensive information.

Detlev Ganten
University of Heidelberg
Heidelberg, FRG

Donald Pfaff
Rockefeller University
New York, USA

Contents

Neurobiology of REM Sleep

A Possible Role for Dopamine

GÜNTER STOCK[1]

Contents

1 Introduction

As early as 1867 Wilhelm Griesinger reported on eye movements not related to vision but related to fantasies. These eye movements occurred both during the transition from wakefulness to sleep and during dreaming. From these observations he concluded that sleep was not a passive but rather an active state. It was another 86 years before

1 I. Physiologisches Institut, Universität Heidelberg, Im Neuenheimer Feld, D-6900 Heidelberg, FRG

Aserinsky and Kleitman (1953) discovered that sleep is not a homogeneous process but is organized in rhythmically occurring cycles of different stages, each of which is characterized by specific behavioral, electrophysiologic, autonomic, and endocrine changes. Since then, sleep was grossly divided into nonREM sleep, also called slow wave sleep (SWS), and REM sleep, also called desynchronized sleep (DS), D sleep, and paradoxical sleep (PS).

Both types of sleep, i.e., nonREM sleep and REM sleep, evolved at different times during the phylogeny of mammals and interestingly enough also in birds. It is currently held that nonREM sleep probably evolved 180 million years ago in animals which are considered true mammals. REM sleep, on the other hand, evolved 50 million years later (Allison and Twyver 1970). Probably for this reason and because of the fact that REM sleep is prominent in immature mammals (Roffwarg et al. 1966) the functions of REM sleep have been hypothesized to accompany and probably be causally related to brain maturation. In the adult brain REM sleep could serve as a means by which the organism could excercise more adaptivity and more plasticity of innate behavior (Jouvet 1978).

The present paper aims (a) to review available data which characterize REM sleep in mammals and (b) to discuss possible mechanisms underlying the triggering and maintenance of REM sleep. No attempt will be made, however, to cover the expanding field of peptidergic mechanisms during sleep, since this will be dealt with in a separate chapter of this volume. No attempt will be made either to discuss in detail the available data on lesion experiments and on the pharmacology of sleep, several extensive reviews having dealt earlier with these topics (King 1971; Moruzzi 1972; Jouvet 1972, 1977; Morgane and Stern 1974). This review rather attempts to mainly summarize those data which can be obtained from animals (and probably from man) under physiologic conditions. Therefore spontaneously occurring electrophysiologic or biochemical events will be considered primarily. In the last section data which point to a possible involvement of dopamine (DA) in the generation of REM sleep will be shown since this aspect has not yet been fully covered in recent reviews.

The following abbreviations appear in the text:

ACH: acetylcholine; ACTH: adrenocorticotrophic hormone; α-MT: DL-α-methyltyrosine methylester HCl; CSF: cerebrospinal fluid; DA: dopamine; DA-β-hydroxylase: dopamine-β-hydroxylase; Dopa, L-Dopa: 3,4 dihydroxyphenylalanine, precursor of DA and NA; dopac: 3,4 dihydroxyphenylacetic acid, metabolite of DA; DS: desynchronized sleep; EEG: electroencephalogram; EMG: electromyogram; FTG: gigantocellular tegmental field; GABA: gamma-aminobutyric acid; GBL: gammabutyrolactone; GH: growth hormone; GHBA: gammahydroxybutyric acid; 5-HIAA: 5-hydroxyindole-3-acetic acid, metabolite of 5-HT; 5-HT: 5-hydroxytryptamine, serotonin; 5,6-HT: 5,6-dihydroxytryptamine; 5-HTP: 5-Hydroxytryptophane, precursor of 5-HT; HVA: homovanillic acid, metabolite of DA; LC: locus coeruleus; NA: noradrenaline; NSD 1015: metahydroxybenzylhydrazine-HCl; 6-OHDA; 6-hydroxydopamine; PCPA: para-chlorophenylalanine; PGO-waves: ponto-geniculo-occipital waves; PRL: prolactin; PS: paradoxical sleep; REM: rapid eye movements; SWS: slow wave sleep; TSH: thyroid-stimulating hormone

2 Phenomenology of REM Sleep

2.1 Electroencephalographic Findings

The electroencephalographic features of REM sleep as first described in humans by Dement and Kleitman (1957) and by Dement (1958) in cats are characterized by EEG desynchrony. This is part of the so-called tonic phenomena during REM sleep which were differentiated from phasic events for the first time by Moruzzi (1963). In addition to cortical desynchrony, EEGs from amygdala, caudate nucleus and ventral hippocampus are activated as well, whereas dorsal hippocampus shows a regular theta rhythm (4–6 Hz). Phasic electroencephalographic events related to REM sleep are found in cats; according to the loci from which they are recorded they are designated ponto-geniculo-occipital (PGO) waves. PGO waves are monophasic high voltage waves preceding and accompanying REM sleep whereas they are very seldom observed during nonREM sleep. During REM sleep PGO bursts (interval between individual PGO waves less than 100 ms) rather than single PGO waves are recorded. PGO activity occurs at a rather constant daily rate of 13,000±1,500 in the cat (Buguet 1969). They were first described by Jouvet and Michel (1959) to occur in the pontine reticular formation, by Mikiten et al. (1961) in the lateral geniculate body and by Mouret et al. (1963) in the occipital cortex. In humans no PGO activity or other forms of their representation were observed.

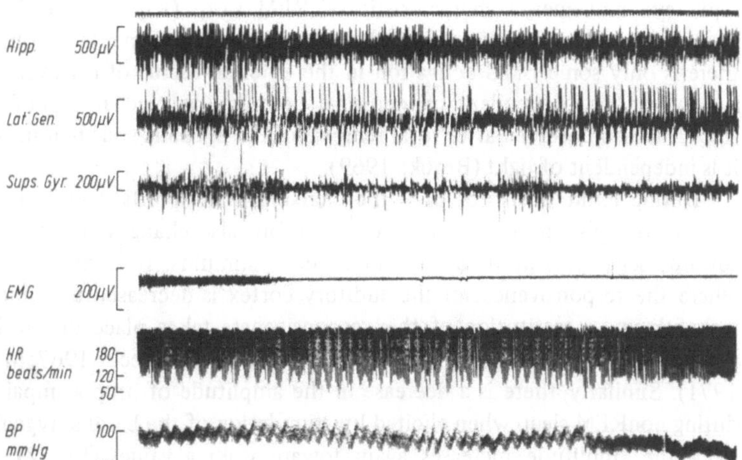

Fig. 1. Transition from slow wave sleep (nonREM sleep) to REM sleep in unrestrained cat. Electroencephalogram of ventral hippocampus *(Hipp)*, lateral geniculate body *(Lat. Gen.)*, and Suprasylvanian cortex *(Sups. Gyr.)* are shown, as well as the electromyogram of dorsal neck muscle *(EMG)* heart rate (*HR*, beats/min) measured as the duration of the interval between two heart beats and arterial blood pressure (*BP*, mmHg). The time scale is given as seconds in the first line. Concomitantly with the appearance of PGO waves there is a short-lasting increase in arterial pressure. Moreover there is a marked variation in heart rate and a decrease in respiratory arrhythmia during REM sleep compared to nonREM sleep

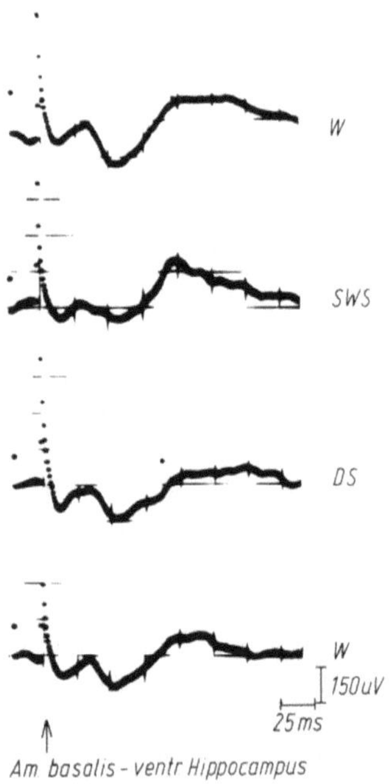

Fig. 2. Evoked potentials recorded within ventral hippocampus following stimulation of the ipsilateral basal amygdala (Am. basalis) at 0.3 ms and 0.1 mA. Forty impulses were averaged and displayed on an oscilloscope during consecutively occurring stages of wakefulness *(W)* nonREM, slow-wave sleep *(SWS)*, REM sleep, desynchronized sleep *(DS)* and *W* again

PGO activity coincides with or precedes another phasic phenomenon, namely the rapid eye movements characteristic of REM sleep (Kiyono and Jeannerod 1967). They are, however, also recorded when no gross eye movements are observed but when there is only some EMG activation in the external rectus of the eyeball. PGO activity is however, not dependent on neuronal or muscular mechanisms of eye movements since it is also observed after total removal of the eyeballs and their muscles. Moreover, it is independent of light (Brooks 1969).

During REM sleep not only the characteristic, spontaneously appearing electroencephalographic patterns can be observed but also changes of cortical responsiveness towards sensory stimuli such as an acoustic stimulus. In contrast to nonREM sleep where the responsiveness of the auditory cortex is decreased to stimuli compared to wakefulness, a restitution of this responsiveness takes place during REM sleep and reaches a level comparable to that of wakefulness (Berlucchi 1967; Murphy and Starr 1971). Similarly there is a decrease in the amplitude of hippocampal field potentials during nonREM sleep when elicited by stimulation of the basal amygdala. During REM sleep the amplitude increases again toward waking values (Fig. 2). Hence, marked differences in electrical excitability of cortical and limbic structures are observed during nonREM sleep compared to wakefulness and REM sleep.

2.2 Electromyographic Findings

During REM sleep there is a decrease in electromyographic activity of cervical, submental, and facial muscles as well as of the muscles of the extremities (Berger 1961; Bliwise et al. 1974; Jouvet and Michel 1959) leading to complete muscle atonia which in the cat can be rather abrupt (Gschwend 1970). This atonia is paralleled by a depression of somatic monosynaptic as well as of polysynaptic reflexes both at the level of the brainstem and spinal cord (Baldissera et al. 1966; Pompeiano 1967; Chase et al. 1968), the inhibition being due to pathways descending in the ventrolateral funiculi of the spinal cord (Gassel et al. 1964). According to Gassel and Pompeiano (1965) and Kubota and Tanaka (1968) the muscular hypotonia apart from inhibition of α-motoneurons could also result from an inhibition of γ-motoneurons by interfering with fusimotor functions. A similar conclusion was reached by Hodes and Dement (1964), and by Beck et al. (1973) in man.

From the findings of a decrease in the discharge rate of motoneurons, Morales and Chase (1978), and Nakamura et al. (1978) designed experiments in which they subsequently showed that in REM sleep both trigeminal motoneurons (innervating jaw-closer muscles) and lumbar motoneurons were hyperpolarized by 3–10 mV. Moreover, subthreshold synaptic activity decreased in jaw-closer motoneurons (Nakamura et al. 1978). Altogether, these findings of a tonic membrane-hyperpolarization of motoneurons at the level of the brainstem as well as at the level of the spinal cord can explain the tonic somatomuscular effects during REM sleep. The hyperpolarization itself seems to be due to a marked postsynaptic inhibition (Morales and Chase 1980).

Besides phasic electrocorticographic phenomena there are phasic muscular activations which characterize REM sleep both in man and in cat, namely phasically occurring conjugated eye movements or rapid eye movements (REM) (Aserinsky and Kleitman 1953; Dement 1958; Jouvet 1962). Moreover, there are muscle twitches during these REM phases which affect vibrissae, ears, jaw, limbs, and tail. They occur during bursts of REM during which there is an additional phasic inhibition of residual muscle activity and reflex inhibition (Gassel et al. 1964). These muscle twitches are more frequent in distal than in proximal muscles and more prominent in flexor than in extensor muscles. They depend on phasic, short-lasting but powerful excitatory volleys arising from supraspinal areas descending through the dorsolateral funiculi and overwhelming the ongoing inhibitory influence upon spinal mono- and polysynaptic reflexes (Gassel et al. 1964). Although these phasic phenomena can be rather intense they cannot overcome the ongoing postural atonia. The anatomic substrate of this atonia was shown to be located bilaterally in the medial parts of the locus coeruleus α and perilocus coeruleus α (Pompeiano 1970; Sastre 1978; Sakai 1980). Fibers of this system project to the spinal cord via the tegmentoreticular tract. This has been proven both by lesion techniques as well as by single unit recordings (cf. Sakai 1980).

2.3 Autonomic Nervous System

2.3.1 Hemodynamic Data

A tonic reduction in mean arterial pressure both in man (Snyder et al. 1964; Bristow et al. 1969; Coccagna et al. 1971; Sleight, this volume) and in the cat (Mancia et al. 1970; Stock et al. 1978) has been described during REM sleep. This decrease is mainly due to a decrease in total peripheral resistance (Bristow et al. 1969). In the cat regional changes in blood flow during REM sleep have been followed in greater detail. According to Mancia et al. (1970) cardiac output decreases slightly whereas total peripheral resistance decreases more markedly. There is an increase in mesenteric and renal conductance (inverse of resistance), and a decrease in iliac blood flow especially in the red muscle blood flow, though the blood flow in the white muscles remains unchanged (Reis et al. 1969). Superimposed upon these tonic changes are phasic events during REM sleep: an increase in arterial blood pressure, an increase in iliac and occasionally mesenteric resistance. Furthermore, in man there is evidence for the existence of a superimposed circadian rhythm due to which arterial blood pressure values are generally lower during the first hours of sleep and gradually increase towards the second part of the night (Coccagna et al. 1971; Khatri and Freis 1967; Millar-Craig et al. 1978). In cats there is a trend toward lower values in arterial pressure if consecutive REM episodes are followed. This holds true both for spontaneously occurring REM sleep as well as for REM sleep following injection of 5-Hydroxytryptophane (5-HTP) in cats previously depleted of serotonin (5-HT) by application of para-chlorophenylalanine. At the onset of REM sleep there is a short-lasting increase in arterial pressure in humans (Snyder et al. 1964) and in cats (Candia et al. 1962) which parallels the onset of PGO activity (Fig. 1).

2.3.2 Heart Rate

Heart rate decreases during REM sleep compared to wakefulness have been reported both in man (Snyder et al. 1964; Smyth et al. 1969) and in cat (Jouvet et al. 1960; Stock et al. 1978). The most marked finding, however, consists of an enhanced variability of heart rate during REM sleep, with both large increases as well as large decreases. Moreover, there is marked variation also concerning the respiratory arrhythmia. Respiratory arrhythmia is characterized by a decrease in heart rate during expiration and an increase in heart rate during inspiration. This type of arrhythmia is attenuated during approximately half of the period of REM sleep (Stock et al. 1980, see also Figs 1, 3). This again underlines the variability of phasic heart rate changes during REM periods. There have been efforts to clarify the mechanisms of the tonic decrease in heart rate during REM sleep as well as that of the phasic changes. Baust and Bohnert (1969) concluded from their studies in cats that the tonic decrease in heart rate seen in nonREM sleep is mainly due to an increase in vagal tone of the heart whereas a further fall during REM periods is mainly brought about by sympathetic withdrawal. The phasically occurring tachycardia during REM sleep is probably due to vagal withdrawal (cf. Fig. 3) whereas bradycardia is due to an increase in vagal and a decrease in sympathetic tone. These conclusions are corroborated by results obtained with chronic recordings of vagel (Leichnetz 1972) and sympathetic nerve activities (Baust et al. 1968; Kirchner 1973).

In addition to spontaneously occurring heart rate changes there are changes in cardiovascular regulation, e.g., cardiovascular reflexes. In man it has been shown that there is an increase in the sensitivity of the baroreceptor reflex which is more pronounced in REM sleep than during nonREM sleep (Smyth et al. 1969; Sleight, this volume).

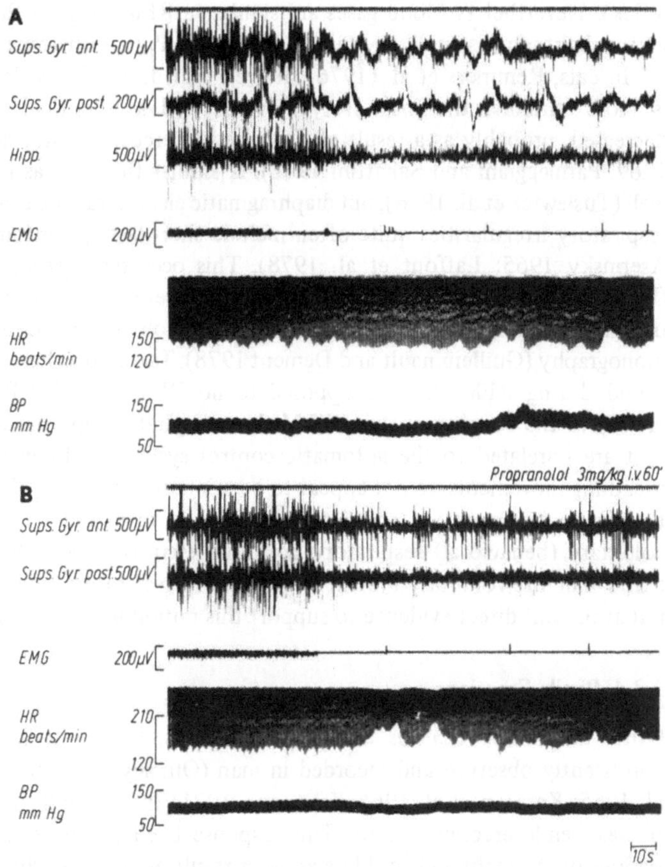

Fig. 3. *Panel A:* Recordings are similar to Fig. 1 and were performed in an unrestrained cat. Note that there is a marked variability in heart rate and a decrease in respiratory arrhythmia during REM sleep. Both the tachycardiac periods during REM sleep and the decrease in respiratory arrhythmia are unaffected by pretreatment with a high dose of a beta receptor blocking agent, propranolol, 3 mg/kg given i.v. 60 min before recording the REM period *(panel B)*

2.3.3 Respiration

As early as 1923 Mac Williams noted that "disturbed sleep" (REM sleep) in contrast to "undisturbed sleep" (nonREM sleep) was characterized not only by increases in heart rate and blood pressure but also by increases in respiratory rate. This was presumably the first description of changes within the autonomic nervous system which were regarded as the autonomic counterpart of dreaming. Since then it has consistently been shown that in humans respiratory activity is slightly increased during REM sleep. Moreover, there is a greater variability in respiratory rate (Snyder et al. 1964). This latter finding matches those observed in dogs where respiration has been shown to be highly irregular. In parallel there is a decrease in responsiveness toward CO_2 (Phillipson et al. 1976) and toward vagal stimuli (e.g., the classical Hering-Breuer reflex). Nevertheless blood gases are similar in REM sleep to those in nonREM sleep with a slight increase in P_{aCO_2} and a decrease in P_{aO_2} (Bristow et al. 1969).

In cats, Remmers et al. (1976) showed that during REM sleep respiratory volume is more variable, the inspiratory flow as well as the tidal volume being slightly decreased, probably as a result of decreased intercostal motor activity in cats (Duron 1969; Parmeggiani and Sabattini 1972). A similar finding was reported for humans as well (Tusiewicz et al. 1976), but diaphragmatic electrograms are not affected. In humans respiratory irregularities quite often include short-lasting central apneas (Bülow 1963; Aserinsky 1965; Laffont et al. 1978). This occurrence seems to increase with age (Webb 1974). These respiratory disorders, however, become clinically significant in patients suffering from sleep disorders. This problem has been discussed in a recent monography (Guilleminault and Dement 1978). Taken together, the respiratory changes found during REM sleep as opposed to nonREM sleep led Phillipson and Sullivan (1978) to the conclusion that "REM sleep respiration appears to be driven by inputs that are unrelated to the automatic control system, and considered in this context, breathing movements do not appear to be primarily 'ventilatory' in origin or purpose ... they are reminiscent of ventilatory activity in man under conditions in which the voluntary (behavioral) respiratory control system is active. Therefore, the possibility exists that the voluntary system is involved in respiratory control during REM sleep, but at present direct evidence to support this notion is still lacking."

2.3.4 Penile Erections

Other autonomic changes are penile erections. During REM episodes it has been consistently observed and recorded in man (Ohlmeyer and Brilmayer 1947; Fisher et al. 1965; Karacan et al. 1966; Jovanovic 1971) that periodically, i.e., parallel to REM phases penile erections occur. This response becomes more marked during the late periods of the night and could be another result of circadian rhythm.

2.3.5 Endocrine Changes

Anterior pituary hormone secretion (GH, ACTH, TSH) typically follows a circadian rhythm exhibiting a maximum during sleep (Takahashi 1979). For human plasma prolactin (PRL) it has been clearly established that, unlike the other hormones, a significant association exists between REM episodes and nadirs in PRL secretion. The secretion of PRL increases and peaks during subsequently occurring nonREM phases (Parker et al. 1974). Since DA, released from the hypothalamic DA system, has been shown to be inhibitory to the secretion of PRL (Kamberi et al. 1971; Kleinberg et al. 1971) this evidence supports an involvement of DA in the executive mechanisms of REM sleep (vide infra).

3 Single Unit Activity During REM Sleep

According to the classic lesion experiments (cf. Jouvet 1972; Moruzzi 1972) and from the results obtained in cats with chronic prepontine transections ("pontine cats") the

REM-generating structures were localized in brainstem structures lying between the spinal cord and anterior pons. For that reason most of the single unit recordings were performed in this area. The principal aim of these studies was to find cells which specifically initiate different episodes of sleep characterized by an increase or decrease in activity. Therefore these cells could be regarded as executive cells. In addition to tonically active cells during REM sleep, units should be found which accompany or even precede the phasically occurring electroencephalographic, autonomic or motoric changes within the REM period. Both types of units should act not only on descending (e.g., muscle atonia, see above) but also on ascending systems, described by Moruzzi (1972), ultimately altering cortical and subcortical cell activity. In the following section a short summary of unit recordings will be given starting at the cortical level and ending at the brainstem level, from which "pacemaker activity" with regard to REM sleep should be expected, if the working hypothesis is correct.

3.1 Cortex

The first recording of cortical single unit activity in unrestrained cats was performed by Hubel (1959) showing two types of unitary activity in the striate cortex: one type becoming regular in firing during arousal compared to bursting activity during nonREM sleep, and one type becoming inactive during arousal. No information concerning REM sleep was given in this early paper. It remained for Evarts (1962) to show an increase in the firing rate of units within the visual cortex of the cat during REM sleep. A similar finding was obtained also for neurons of the lateral geniculate body (Sakakura 1968) during REM sleep. Noda and Adey (1970a, b, c) reported on units within the association cortex of the cat. A decrease in firing rate was observed during nonREM sleep which increased again during transition to REM sleep. In contrast to pyramidal tract neurons, where bursting patterns of unit firing were common during REM sleep (Evarts 1964), neurons of the association cortex had a frequency comparable to wakefulness. During REM sleep, however, longer silent periods were observed. This has been interpreted as evidence for the lack of correlation of the firing rate with phasic motor activities during REM sleep (Noda and Adey 1970c).

In addition to the pattern of firing observed by Evarts (1974), Hobson and McCarley (1971) reported that cells within the posterolateral cortex of the cat increase their firing rate from wakefulness to nonREM sleep and even further during REM sleep. These changes often preceded the usual electroencephalographic characteristics from several seconds up to 1 min.

3.2 Thalamus

As already mentioned in the preceding section, Sakakura (1968) recorded from units within the lateral geniculate body and found increased activity during REM sleep which was characterized by bursts of spikes correlated with rapid eye movements. There was a short-lasting inhibition of discharge immediately before PGO activity was recorded and an increase during the PGO activity. This firing pattern was independent from retinal inputs as was demonstrated by Sakakura and Iwama (1967) in enucleate cats; hence it was concluded that the pattern depended on brainstem or cortical influences rather than on exteroceptive input.

Within the nucleus reticularis thalami as well as within the nucleus ventralis thalami, unit activity decreased during nonREM sleep and increased again during REM sleep (Mukhametov et al. 1970; Lamarre et al. 1971).

3.3 Limbic Structures

Few data are available concerning unitary activity within the limbic system of unrestrained animals although these nuclei have repeatedly been implicated in arousal mechanisms (Pribram and McGuiness 1975). Noda et al. (1969) recorded units within the *hippocampus* and found high discharge rates during active waking and REM sleep which were not directly related to phasic events and lower rates during quiet waking and nonREM sleep. Mink et al. (1967) reported on six cells within dorsal hippocampus which had higher discharge rates during nonREM sleep than during REM sleep. Within the *basal amygdala* three types of units have been found to date. One type has higher discharge rates during nonREM sleep than during wakefulness or REM sleep. Jacobs and McGinty (1971) have reported that the majority of units (39 out of 61) are of this type. According to this study only 10 out of 61 units responded in the way in which most of the higher brainstem units seem to react, i.e., an increased firing rate during arousal and REM sleep compared to nonREM sleep. O'Keefe and Bouma (1969) reported on seven units (type 3) within the basal and lateral part of the amygdala in unrestrained cats which did not react to different states of arousal but rather toward "complex sensory stimuli" (e.g., simulated bird calls or presentation of

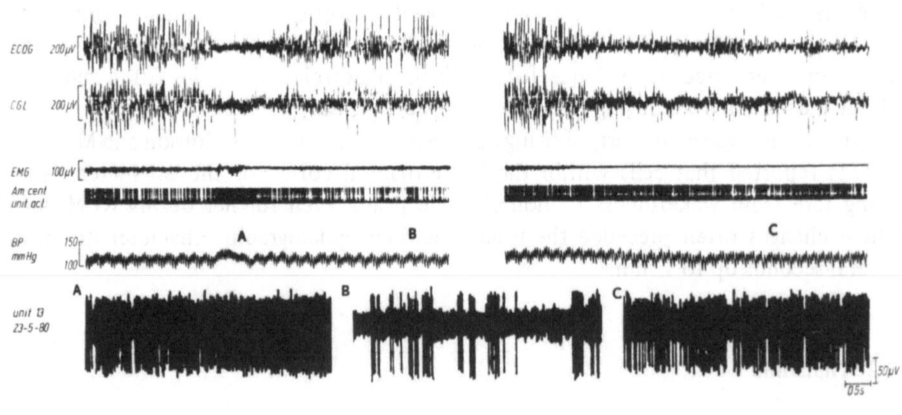

Fig. 4. Concomitant recordings of cortical *(ECOG)* and geniculate *(CGL)* EEGs, *EMG* activity, amygdaloid unit activity, shown as triggered pulses (Am. cent. unit act.), and arterial blood pressure during spontaneous arousal *(A)*, nonREM *(B)* and REM sleep *(C)* in an unrestrained cat. Oscilloscope tracings of the amygdaloid unit are given according to the periods denoted *A*, *B*, *C*

a mouse). In our own experiments we recorded exclusively from the *central part of the amygdala* in unrestrained cats. These units decreased their firing during nonREM sleep compared to arousal or quiet wakefulness and increased again during REM sleep with occasional silent periods (Fig. 4). The changes in the firing rate during REM sleep and arousal were not correlated to EMG activation of the dorsal neck muscles and they had a different time course than electroencephalographic parameters.

3.4 Hypothalamus

Findlay and Hayward (1969) identified hypothalamic units in the rabbit characterized by marked increases in their discharge rate during REM sleep compared to nonREM sleep and wakefulness. Oomura et al. (1969), however, found units within the ventrolateral part of the hypothalamus which had higher discharge rates during nonREM sleep than during REM sleep or during arousal.

3.5 Reticular Formation

The *pontine nucleus giganto cellularis* (the gigantocellular tegmental field, FTG, according to Berman 1968) occupies a large portion of the reticular formation of the brainstem and also includes the nucleus reticularis pontis and oralis and hence has extensive afferent and efferent connections which may be of functional importance. (For a detailed discussion of functional implications of the morphology of this area see a recent review (Steriade and Hobson 1976). In this area Siegel et al. (1977, 1979) found three different cell types: Type I cells (28% of the investigated population) which were only active in association with muscular movements; type II cells (15%) with high spontaneous activity (> four impulses/s) both during waking and sleep; and type III cells with a somewhat lower firing rate, discharging in bursts during waking movements and REM sleep. Siegel (1979) concluded that units within the gigantocellular nucleus are mainly concerned with motor output. On the contrary other investigators (Pompeiano and Hoshino 1976; McCarley and Hobson 1971; Hobson et al. 1974) have evidence in favor of the existence of gigantocellular cells which are selective for REM sleep. This was concluded from autocorrelational analyses (McCarley and Hobson 1965) and cross-correlation techniques (Pivik et al. 1977), insofar as these cells were shown to discharge in clusters beginning 50–150 ms before the onset of eye movements. Since such a correlation was not obtained for eye movements in awake animals it was concluded that this type of FTG cells was causally involved in eye movements generation during REM sleep. In a recent series of experiments cells within the FTG region could be identified which are situated in apposition to the brachium conjunctivum and discharge in stereotyped bursts of about six spikes, 12 ms before the onset of PGO waves in the ipsilateral geniculate body. This firing pattern has led to the assumption that these units are output neurons within the neuronal system which generates PGO waves (McCarley 1978; Nelson et al. 1980) and probably other phasic events of REM sleep. Sakai (1980) found at three different locations units: (a) within an area which begins rostrally at the end of the caudal end of the oculomotor nucleus and is called area X; (b) within the nucleus latero-dorsalis tegmenti;

and (c) within the nucleus parabrachialis lateralis, which discharged 5–25 ms prior to the onset of PGO waves within the lateral geniculate body. Rapid eye movements, on the other hand, which are correlated with these PGO waves seem to depend on activation of neurons of the medial and descending vestibular nucleus (Bizzi et al. 1964; Pompeiano 1980). McCarley et al. (1980) have presented evidence that the discharge of PGO-burst neurons is not only linked to laterality of PGO waves within the geniculate body but also to laterality of eye movements during REM sleep. Netick et al. (1977) recorded from the *medullary gigantocellular nucleus* which is a caudal continuation of the pontine gigantocellular area. Units in this area were either silent or were intermittently active during deep nonREM sleep which precedes REM sleep or began to discharge at a gradually accelerating rate with the occurrence of REM sleep. Thereafter they discharge tonically, with modulation, throughout REM sleep and cease at the end of REM sleep. From this behavior these cells were regarded as prime candidates for the generation of REM sleep (Netick et al. 1977). However, Siegel (1979), using unrestrained cats as opposed to Netick et al. (1977), only recorded from cells in this area which were related to motor output and were not dependent on REM sleep per se.

3.6 Locus Coeruleus

Cells, both within locus coeruleus and subcoeruleus were shown to have lower discharge rates during REM sleep than during nonREM sleep and wakefulness (Hobson et al. 1975; Chu and Bloom 1973). According to Chu and Bloom (1974) these cells are noradrenaline (NA)-containing neurons being reciprocally connected (both anatomically and functionally) with the FTG cells (McCarley 1978). Cells of this type have been called D-off cells. Besides this cell type, D-on cells could also be found in this area intermingled with D-off cells. Because of methodological limitations, a clear distinction has not been made which allows us to unequivocally state which (D-off or D-on) cells are, in fact, of noradrenergic nature.

3.7 Substantia Nigra

Up to now six units located in the pars compacta of the substantia nigra have been found to have higher discharge rates during wakefulness and REM sleep compared to nonREM sleep in unrestrained cats (Fig. 5). As yet it has not been possible to show if these units were dopaminergic in nature.

3.8 Raphe Nuclei

Dorsal raphe nucleus: In freely moving cats cells decrease their discharge rate during REM sleep (D-off cells) compared both to nonREM sleep and wakefulness, and increase the rate again a few sec before the end of REM sleep (McGinty 1973; McGinty and Harper 1976; Trulson and Jacobs 1979). These units are inversely correlated

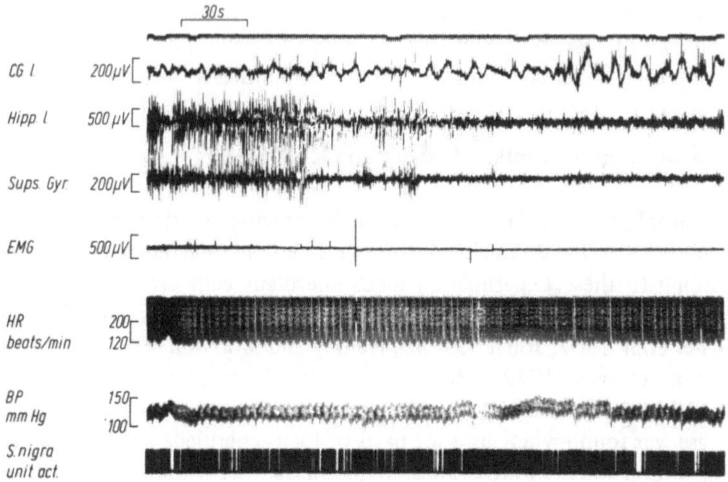

Fig. 5. Transition from nonREM to REM sleep in an unrestrained cat. Recordings as in Fig. 1. Unit activity within the pars compacta of the substantia nigra (S. nigra unit act.) is shown additionally. Neuronal activity increases during REM sleep. (The same nigral unit is shown in Figs. 12 and 13)

with the occurrence of PGO spikes and have been thought to be serotoninergic (Trulson and Jacobs 1978, 1979).

Median and magnus raphe nuclei. Cells in this area increase their firing rate from 5.7/s to 15.9/s when cats pass from nonREM sleep to REM sleep (Sheu et al. 1974). Since a group of raphe magnus neurons increased their discharge rate before the onset and decreased their discharge rate with cessation of hippocampal theta rhythm, it was concluded that the generation and maintenance of theta rhythm observed in the dorsal hippocampus during REM sleep was dependent on these neurons.

From the data obtained with the method of extracellular recording of unitary activity within the brainstem McCarely and Hobson (1975) conclude that REM sleep is generated within a neuronal loop originating in the area of the pontine brainstem where the giant cells of the tegmental field are located. Because of the fact that these neurons have extensive efferent and afferent projections (Brodal 1957; Scheibel et al. 1955) they could in fact serve a regulatory role in complex behavioral patterns, such as sleep-wakefulness behavior. And indeed, FTG cells discharge rather selectively during REM sleep as was outlined above. Since dorsal raphe and locus coeruleus (LC) neurons discharge in a reciprocal manner, i.e., they decrease their discharge rate before the onset of REM sleep, these authors proposed a model which states that FTG neurons are actively inhibited by LC and/or dorsal raphe neurons during wakefulness and nonREM sleep and are disinhibited during REM sleep. This theory of reciprocal interaction between FTG cells and locus coeruleus receives support from anatomic studies which revealed that there are fiber systems connecting the FTG area reciprocally both with locus coeruleus and with dorsal raphe nucleus (Nauta and Kuypers 1958; Chu and Bloom 1974; Conrad et al. 1974). Because of the extensive projection of the FTG cells they could modify the discharge of cells in the terminal regions of the FTG projections. Such a hypothesis could explain why the majority of units in different areas of the brainstem, diencephalon, limbic system, and cortex discharge in a similar pattern during transition from wakefulness to nonREM sleep and REM sleep.

3.9 Commentary to Unitary Activity and REM Sleep

The neurophysiologic data reviewed in the previous sections do not completely agree with observations and conclusions drawn from pharmacologic and lesion experiments. According to these experiments, locus coeruleus cells are essential for inducing and maintaining REM sleep or at least for several of its components like postural atonia (Roussel et al. 1967; Jouvet 1973b; Henley and Morrison 1974; Sastre 1978) and EEG activation (Jouvet 1972). However, as already mentioned by Roussel (1967) the relationship seems not to be a simple one; an increase in REM sleep rather than a decrease was found when anterior parts of locus coeruleus were ablated (Roussel 1967; Petitjean and Jouvet 1970). These data contrast with those obtained with recordings of single unit activities.

Lesion experiments create severe problems such as specificity and extent of the lesion, disturbance of regional blood circulation by surgical trauma and local edema and disruption of traversing pathways. On the other hand, recordings of single unit activity in unrestrained animals also offer a variety of problems. It is apparently rather difficult to avoid a methodological bias, since it is much easier and much more likely that recordings are made from big cells than from small ones. This is of special interest since monoaminergic neurons are small neurons. Moreover, not all studies performed so far can exclude the fact that the discharge rate of the neurons is correlated with physiologic and behavioral parameters, not being tested for methodological and other reasons. They might reflect peripheral variations, either afferent or efferent to the reticular formation instead of being generator neurons, stricto sensu, for REM sleep. Indeed it cannot be excluded that the unit activity could be related, e.g., to hippocampal theta rhythm, discrete muscle twitches or cardiovascular changes which are all common both to waking and REM sleep. Hence the function of the units upto now related to REM sleep might very well extend beyond REM sleep (Vertes 1977; Siegel 1979). In addition, it cannot be totally excluded that FTG neurons which have a key position in the above-mentioned theory of REM sleep generation, might very well receive a descending input from neurons located in diencephalic, limbic or cortical areas and would hence be activated secondarily instead of being primarily involved in the REM sleep-generating processes. Needless to say that extracellular recordings from units thought to be responsible for the periodical recurrence of the sleep-waking cycle, cannot give conclusive evidence for pacemaker activity of cells. From these considerations it is evident that our knowledge concerning REM sleep generation is still fragmentary.

4 Role of Monoamines in REM Sleep Generation

Apart from recordings of unitary activity and apart from spontaneously occurring changes in brain metabolites the most important step toward an understanding of the generation of sleep, especially of REM sleep, was the theory developed by Jouvet and his colleagues which greatly reinforced sleep research.

This theory was probably the first approach to integrate both the notion of Moruzzi that sleep is an active process, an idea that evolved from the earlier lesion and stimulation experiments (cf. Moruzzi 1972) as well as the results of more recent histochemical and neurochemical data. It is not intended to give a detailed review on this theory (cf. Jouvet 1972, 1977), however, the available data indicating a role for acetylcholine and monoamines involved in the REM sleep generation will be briefly summarized. More recent evidence for an involvement of DA will also be discussed.

4.1 Biochemical Correlates of REM Sleep

During REM sleep there is an increase in acetylcholine (ACH) measured in cortical cups in the cat (Jasper and Tessier 1971) as well as in the effluent of a push-pull cannula inserted into the head of the caudate nucleus (Gadea-Ciria et al. 1973). Although this finding alone does not prove that cholinergic mechanisms are actively involved in REM sleep generation the available pharmacologic data in both animals and men (cf. Sitaram and Gillin 1979) make it likely that ACH indeed actively takes part in generating REM sleep as was proposed by Hernandez-Péon (1962, 1963), Jouvet (1965, 1972), Steriade and Hobson (1976). Little data is available concerning spontaneously occurring changes in the metabolism of biogenic amines during REM sleep. Kovacevic and Radulovacki (1976) reported an increase in hippocampal serotonin turnover during nonREM sleep but they did not report on similar changes during REM sleep. Wyatt et al. (1974) observed an increase in 5-hydroxyindole acetic acid (5-HIAA), a metabolite of serotonin (5-HT) during nonREM sleep and a decrease during REM sleep when measured in the CSF obtained from the frontal horn of the lateral ventricle in patients with presenile dementia. Therefore, these data corroborate the idea (vide infra, Jouvet 1972, 1973) that 5-HT turnover might be increased during nonREM sleep and hence may be actively involved in the executive mechanisms of nonREM sleep. On the other hand, during REM sleep 5-HT-containing neurons should be less active as reflected by a decrease in 5-HIAA.

Mouret et al. (1978) reported on several groups of patients, two of which showed characteristic variations in REM sleep and DA metabolism. In patients suffering from hypersomnia characterized by a rather specific increase in REM sleep there was a concomitant increase in lumbar homovanillic acid (HVA). Patients suffering from a rather selective decrease in REM sleep also had low HVA levels determined at the lumbar level. Cespuglio et al. (1979) using the voltametric technique reported that serotonin-turnover decreased within the cortex during REM sleep whereas there was an increase in catecholamine turnover within the caudate nucleus compared to nonREM sleep. Since the caudate nucleus contains mainly DA rather than noradrenaline (NA), the finding might indicate that DA is actively involved in REM sleep.

4.2 Neuropharmacology of REM Sleep

Using pharmacologic, neurotoxic or surgical manipulation of monoamine functions by interfering with synthesis, storage, or release of amines, the activity within ascending

5-HT neurons originating in the nuclei of the raphe system (Andén et al. 1965; Dahlström and Fuxe 1964, 1965) has been shown to be involved in *priming* REM sleep besides its executive effects in slow wave sleep (Jouvet 1973). This conclusion is based mainly on the finding that after synthesis inhibition of 5-HT by application of parachlorophenylalanine (PCPA) there is a decrease and finally a disappearance of REM sleep when nonREM sleep is decreased below 15% of its "normal daily" amount (Koella et al. 1968; Jouvet 1969). On the other hand, if the synthesis of 5-HT is restored by application of 5-HTP (5-hydroxytryptophane) there is an increase in REM sleep which exceeds the duration in control animals (Pujol et al. 1971).

Another argument is the finding that intraventricular injection of 5,6-dihydroxytryptamine (5,6-HT), a neurotoxic agent which rather selectively destroys 5-HT neurons, leads both to a reduction in nonREM sleep and to a significant decrease in REM sleep. This is significantly correlated with the decrease in brain content of 5-HT as well as of 5-HIAA (Baumgarten et al. 1972). These data are suggestive for a role of 5-HT in priming REM sleep. There is, however, a report from Ursin (1972) showing that after single doses of PCPA in cat there was a reduction in nonREM sleep ("deep slow wave sleep") but practically no change in REM sleep. Although these data cannot exclude a role for 5-HT in REM sleep generation they nevertheless point to the possibility that different aspects of the complex sleep-waking cycle are affected if the methodological approaches are slightly modified. This means that the dosage, the route of administration and most probably also the time-schedule of drug application in relation to the sleep-waking cycle are major determinants of the experimental results.

According to the "monoaminergic theory", executive mechanisms for REM sleep are located within the pontine reticular formation where ascending NA neurons originate (Dahlström and Fuxe 1965). These neurons, primarily from the locus coeruleus have a widespread distribution characterized by a bifurcation (with long ascending and descending branches) and by the fact that they give off numerous collaterals along their course throughout the brainstem (Lindvall and Björklund 1978). This type of arborization makes it likely that this system could function in a regulatory system for complex patterns as was already mentioned to be characteristic for FTG neurons.

In the following section some of the arguments will be briefly summarized which led to the theory that NA might be a prime candidate for executive mechanisms in REM sleep.

a) Electrolytic lesions of neurons located in the pontine reticular formation which are near or within the locus coeruleus either decrease REM sleep (Buguet 1969) if performed extensively or interfere with postural atonia (subcoeruleus, vide supra) if only parts of this area are destroyed.

b) Injection of 6-OHDA into the region of locus coeruleus destroys NA neurons (Ungerstedt 1969) and consequently suppresses REM sleep (Buguet 1969; Buguet et al. 1970).

c) Reserpine which depletes catecholamine vesicles reduces REM sleep. REM sleep is restored after a few hours if the precursor of NA, L-Dopa, is given to the animals (Matsumoto and Jouvet 1964).

d) If the NA neuron is supplied with a false precursor (α-methyldopa it is assumed that a false transmitter will be synthesized (α-methyl-NA) whereby REM sleep is suppressed (Dusan-Peyrethon et al. 1968).

e) Inhibition of the enzyme DA-β-hydroxylase also inhibits REM sleep (Dusan-Peyrethon and Froment 1968).
f) Injections of α-receptor blocking agents (e.g., phentolamine) were expected (cf. a–e) to decrease REM sleep. However, many investigations have revealed an increase in REM sleep after α-receptor blockade (Putkonen 1979). The discrepancy can be explained if one assumes that in this case presynaptically located NA receptors of the α_2-type are blocked, thereby increasing NA release and leading to an increase in REM sleep. Indeed, α_2-receptor-stimulating agents, such as clonidine, suppressed REM sleep (Putkonen 1979).

From these experimental results Jouvet (1972, 1973 a, 1977) and recently Putkonen (1979) concluded that NA has to be released in order to generate REM sleep. There were, however, reports indicating that the role of NA might be permissive rather than executive in REM sleep. King and Jewett (1971), Hartmann et al. (1971), Wyatt et al. (1971 a) showed that synthesis inhibition of NA, using α-methyl-paratyrosine (α-MT) leads to an increase in REM sleep both in animals and man. This result is contradictory to a stimulatory role of NA and was somewhat unexpected. Kafi et al. (1977) presented evidence that this finding could be due to unspecific, toxic effects of α-MT. There is, however, another experiment which also does not support a stimulatory role for NA in REM sleep. Jones et al. (1977) reexamined the effect of large locus coeruleus lesions and found no marked alteration in REM sleep despite significant reductions in brain NA.

An interesting problem within the monoaminergic theory of sleep is the role of DA. Despite the fact that the specificity of lesion experiments can be questioned and that pharmacologic manipulations of the NA system usually, to a varying extent, also affect DA systems, this group of neurons has not received much attention with regard to REM sleep. In the following section a brief account of this topic will be given.

4.3 Anatomy of the DA System

According to Lindvall and Björklund (1978) DA neurons originate within the tegmental area of the brainstem. The most medially situated cell bodies project to the septum, nucleus accumbens and to the anteromedial frontal cortex (A 10, according to Dahlström and Fuxe 1964). More laterally situated neurons project to the more lateral and caudal forebrain regions such as olfactory tubercle, perirhinal cortex, rostrolateral parts of the neostriatum and anterior cingulate cortex. Even more laterally situated DA cell bodies (within the substantia nigra, pars compacta and pars lateralis) project primarily to the neostriatum (A 9, according to Dahlström and Fuxe 1964). There are also projections to the amygdala; mainly to the central nucleus and to a lesser degree to the basolateral part (Swanson and Hartmann 1975) and lateral part (Fuxe 1965). Only recently it was shown that the same nigral DA axon which makes synaptic contacts within the neostriatum might also give off collaterals to the globus pallidus (Nobin and Björklund 1973). Within the pars reticulata only scattered DA cells are found. The arrangement of cell bodies and terminal projection areas is reminiscent of some kind of a crude topographic organization.

Another interesting feature of these ascending DA neurons is that, upon stimulation, there is dendritic release of DA within substantia nigra, pars compacta and pars reticulata (Jessel et al. 1976; Korf et al. 1976).

Besides these mesencephalic ascending DA fiber systems there are two systems which originate within the diencephalon.

The *tubero-hypophyseal* DA system originates from cell bodies located in the arcuate nucleus and in adjacent parts of the periventricular nucleus (Fuxe 1964; Björklund et al. 1973). The termination of fibers is within the stalk, the neural lobe and the pars intermedia of the hypophysis (Lindvall and Björklund 1978).

The second intradiencephalic DA system, the *incerto-hypothalamic* DA system, is characterized by a very low DA concentration and originates from dorsal and caudal hypothalamic as well as from periventricular cells and projects mainly into the preoptic anterior hypothalamic region (Björklund et al. 1975). Both the tubero-hypophyseal as well as the incerto-hypothalamic system seem to be involved in neuroendocrine regulation (Hökfelt and Fuxe 1972; Björklund et al. 1974). In general, the DA system, in contrast to the NA system, is characterized by a somewhat more specific projection pattern: there is probably less arborization along its pathway and the terminal projection area is less widespread.

4.4 Functional Aspects of the DA System – Possible Involvement in REM Sleep

At present it is accepted that the DA systems are involved in the expression and elicitation of emotional behavior and locomotor activity (Fibinger and Phillips 1979; Ungerstedt 1979). Similarly, DA has been shown to be involved in disturbances of these functions, mainly in regard to Parkinson's disease characterized by hypoactivity (Hornykiewicz 1979) as well as in schizophrenia characterized by hyperactivity of the DA system (Carlsson and Lindqvist 1963; Angrist and Gershon 1970; Randrup and Munkvad 1968; Crow et al. 1978). Crow (1980) suggested that the positive symptoms of schizophrenia (delusions, hallucinations, and thought disorder) are related to a change in DA transmission, whereas negative symptoms (flattening of affect and poverty of speech) are not primarily related to DA mechanisms. In this context Hartmann's (1976) notion is of interest pointing out marked similarities between the physiologic and psychological similarities of REM sleep (including dreams) and schizophrenia. A similar hypothesis was put forward by Jacobs and Trulson (1979) with respect to a 5-HT and DA interaction during dreams, hallucinations, and psychosis. Furthermore, if a cat's brain is lesioned within the subcoeruleus *a* there is no atonia of skeletal muscles (vide supra) characteristic of REM sleep. Hence the animals display a marked psychomotor behavior during REM sleep, called oneric behavior (Jouvet 1979). This shows that during REM sleep locomotor and psychomotor activity occurs if not prevented by the ongoing marked muscle atonia. Though there is at present no proof that DA might be involved in this behavior, such a possibility cannot be excluded especially if one considers the functions of the DA systems in awake subjects. Jones et al. (1973) reported that lesions of the ascending DA fiber systems do not interfere with the electroencephalographic patterns recorded in awake cats but severely impair behavioral arousal which, in turn, will affect psychomotor behavior. This latter experimental finding is in line with the findings in Parkinson's disease where there is a marked impairment of psychomotor functions (e.g., expressionless face, abolition of associated movements). Hence it is hypothesized that DA may be involved in phasic events of REM sleep. Gaillard and Moneme (1977) observed a slight decrease in the duration of dreams and a marked decrease in dreams of aggressive and sexual contents when healthy young females were treated with sulpiride, a DA receptor-blocking agent with preferential action on mesolimbic DA neurons (Bartholini 1976). If the hypothesis holds true, clinical data on patients with severe degeneration of the nigrostriatal and other DA systems should add valuable information. Perret et al. (1979) reported on a patient with an almost total degeneration of the nigrostriatal DA system who had neither deep slow wave sleep nor REM sleep. Patients suffering from Parkinson's disease have sleep disturbances as well. Although it is not generally accepted (Wyatt et al. 1971b, 1976) there are two reports (Bonasegla et al. 1968; Kendel et al. 1972) which indicate that L-Dopa, the precursor of DA alleviates not only the Parkinsonian symptoms but also increases (i.e. restores) REM sleep.

The prevailing discrepancies within the literature concerning sleep disturbances in Parkinson patients probably result from the fact that these patients are regarded as a homogenous group. Mouret (1975) discussed this problem and found two different groups, only one of which had distinct reductions in REM sleep. It is interesting that these patients were the ones who improved most dramatically after L-Dopa therapy with regard to their neurologic disorders. Another piece of evidence in favor of a role for DA in REM sleep can be found in the fact that in a group of normal and retarded children as well as in adults, L-Dopa increased REM sleep (Castaldo et al. 1973).

Although the evidence presented so far is not fully conclusive, one can nevertheless follow this line of argument by assuming that higher doses of L-Dopa result in an increased formation and release of NA thereby activating the waking system and consequently decreasing REM sleep, whereas lower doses result in a preponderant increase in DA release and in an increase in REM sleep. In line with these arguments are the findings reported above showing that the DA-dependent PRL secretion is inhibited during REM sleep suggesting an increase in DA release within the hypothalamic DA system. Similarly, Mouret's data (1978) on increased levels of HVA in lumbar CSF of patients with increased occurrence of REM sleep support this view.

Another approach to the question if and in which way DA might actively be involved in REM sleep is to look for effects of neuroleptics on REM sleep both in man as well as in animals. Neuroleptics which are considered to mainly block DA receptors yield contradictory results. According to Hartmann (1978), most DA blockers do not interfere with REM sleep in humans, with the exception that in normal subjects, chlorpromazine clearly decreased REM sleep in the 2nd, 3rd, and 4th week of administration, but had no effect on REM sleep during the 1st week. On the other hand, Kay et al. (1976) reviewed evidence showing that chlorpromazine increases REM sleep during the initial period of drug application. In rats (Kafi and Gaillard 1976) and cats both spiroperidol and haloperidol have been shown to decrease REM sleep in a dose-dependent way, whereas application of a DA receptor-stimulating drug, apomorphine, apparently also has dose-dependent effects on REM sleep (Kafi and Gaillard 1976; Mereu et al. 1979). In a recent study Cianchetti et al. (1980) showed that i.v. infusions of apomorphine in nonemetic doses (10–15 μg/min) suppressed REM sleep in humans. If one assumes that the doses used primarily affect presynaptically located DA autoreceptors, the data support an active role of DA in REM sleep.

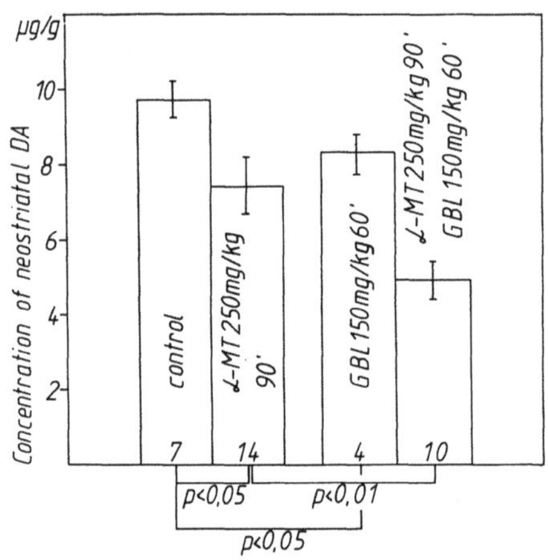

Fig. 6. Measurements of neostriatal *DA* content in Sprague-Dawley rats, under control conditions, and after inhibition of the *DA* synthesis by blockade of the rate-limiting enzyme tyrosine hydroxylase using *α*-methyl paratyrosine (*α-MT*, 250 mg/kg i.p.). After a period of 90 min there is a significant reduction in *DA* levels compared to control values. If gamma-butyrolactone *(GBL)* is injected in low doses, 150 mg/kg i.p. there is a slight reduction in *DA* levels which is enhanced, if *α-MT* has been given prior to the *GBL* injection. Numbers of experiments are given at the bottom of the columns. Statistical analysis was performed using student's *t*-test

Table 1. Measurement of neostriatal DA and Dopa in cats. Cats were decapitated and DA was determined fluorimetrically in untreated cats, in cats which had received low doses of GHBA, 150 mg/kg i.v. and high doses of GHBA, 300 mg/kg 30 min before sacrifice. As can be compared with the data obtained in rats, GHBA, 150 mg/kg (low dose) decreased the neostriatal DA content, whereas the high dose increased neostriatal DA (Stock et al. 1973). If the accumulation of Dopa was measured after inhibition of the enzyme Dopa decarboxylase using NSD, 1015, 100 mg/kg according to Kehr et al. (1972) there was a marked increase in Dopa accumulation if cats had received the low dose of GHBA

Treatment	Neostriatal dopamine (μg/g)	Neostriatal dopa (μg/g)
Control	7.440±0.525 (4)	0.921±0.066 (6)
150 mg/kg GHBA 30 i.v.	7.183±0.727 (6)	1.468±0.0373 (5)
300 mg/kg GHBA 30 i.v.	11.195±1.612 (4)	

Finally, if DA is injected into the brain ventricles of cats, REM sleep is significantly increased (Hartmann and Zwilling 1975) for a period of 2 h after the injection. Although not unequivocally, the results presented so far point to a possible involvement of DA in REM sleep generation. Our own evidence in favor of such an involvement derives from experiments in chronic cats. As a tool we have used gammahydroxybutyrate (GHBA), a naturally occurring metabolite of gamma-aminobutyrate (Doherty et al. 1978) or its lactone-form, gammabutyrolactone (GBL). Both GHBA and GBL have unique properties: whereas DA neurons stop firing when this compound is given in high doses (200–750 mg/kg GBL or 500–2500 mg/kg GHBA) neither Na- nor 5-HT-containing neurons are markedly affected (Stock et al. 1973; Walters et al. 1972, 1973). This leads to a decrease in DA release and hence to a dose-dependent diminuation of the stimulation of presynaptically located DA receptors (Carlsson et al. 1972; Stock et al. 1980b). Consequently a presynaptically located negative feedback

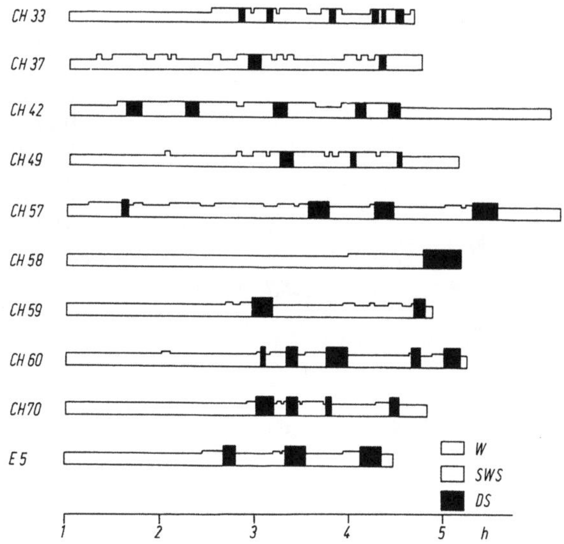

Fig. 7. Sleep-waking cycle in untreated laboratory cats. Wakefulness *(W)*-slow wave sleep, nonREM sleep *(SWS)*, and desynchronized sleep, REM sleep *(DS)* are distinguished. The individual cats usually show the first REM period after a variable latency of 35–180 min from the beginning of the recording session. The time scale is given in hours (h)

mechanism is disinhibited resulting in an increase in intraneuronally stored DA (Andén et al. 1973). There is a distinct anesthesia both in rat and cat as evidenced from behavioral (Roth and Suhr 1970) as well as from electroencephalographic and cardiovascular changes (Stock et al. 1978; 1980a). Quite the opposite holds true when animals receive low doses of GBL or GHBA. In the rat there is a slight decrease in neostriatal DA. If the DA synthesis is inhibited using α-MT, there is an acceleration of the drug-induced decrease in DA content. Such a finding suggests an increase in DA release (Fig. 6). If the rate of DA synthesis is measured in vivo (Kehr et al. 1972) there is a significant increase in the accumulation of dopa, demonstrating that the rate of DA synthesis is increased within the neostriatum of cats after low doses of GHBA (Table 1). These data are corroborated by those of Cheramy (1977) who postulated an increase in DA release after application of GHBA, and by Olianas (1976) who found an increase both of dopa and dopac (a metabolite of DA) in mice after low doses of GBL. From recent electrophysiologic and neuropharmacologic experiments the biphasic dose-dependent effect of GABA-ergic compounds can be tentatively explained by the existence of different GABA receptors, probably located pre- and postsynaptically, with different sensitivities towards GABA-agonistic drugs (Worms and Lloyd 1980).

Behaviorally, low doses of GHBA i.v., will induce REM sleep. The latency is variable within a few seconds up to 10 min following the injection (Laborit 1960; Matsuzaki and Takagi 1967; Stock et al. 1978, 1980a; Figs. 7 and 8). With regard to behavior, cortical and subcortical EEG, EMG, blood pressure, and heart rate spontaneous REM sleep is indistinguishable from GHBA-induced REM sleep (Fig. 9).

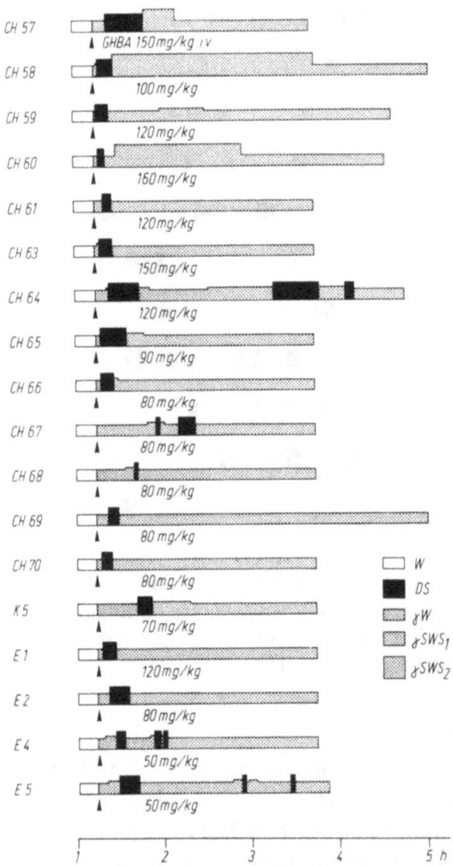

Fig. 8. Arranged as Fig. 7. Cats received low doses of GHBA as indicated under the individual sleep-waking cycles. In addition to wakefulness *(W)* and REM sleep *(DS)* the behavioral states following the GHBA injections were classified as wakefulness *(γ-W)*, slow wave sleep *(γ-SWS_1)* and anesthesia *(γ-SWS_2)*. The time, when GHBA was injected i.v. is denoted by *arrows*. There is some overlap between animals shown in Fig. 7, allowing a direct comparison

Since the stimulatory effects of GHBA on DA neurons parallel the REM-inducing property, this suggests that the increased rate of DA utilization may be causally involved in REM sleep induction. As a first step to verify this hypothesis, cats were treated with para-chlorophenylalanine (PCPA) in order to exclude a possible serotoninergic priming effect on GHBA-induced REM sleep. As can be seen in Fig. 10 cats thus treated did not show spontaneous REM sleep. Nevertheless, GHBA induced REM sleep in the same way as in otherwise untreated cats (Fig. 11). Hence, an effect of GHBA on the elicitation of REM sleep is not likely to act via a serotoninergic mechanism. In a second series experiments with antidopaminergic drugs (haloperidol, pimozide) which block postsynaptically located DA receptors (Andén and Grabowska-Andén 1979) were performed in cats. By this treatment GHBA-induced REM sleep was blocked when the drugs were given in rather high doses (1–2 mg/kg) i.v. before GHBA application. When given in lower (noncataleptic) doses (0.2–0.5 mg/kg) haloperidol failed to block GHBA-induced REM sleep.

This finding could be due to the fact that GHBA does not act via DA when it induces REM sleep. Indeed, it cannot be excluded that GHBA acts via its influence on cerebral glucose utilization (Wolfson et al. 1977) or on cell membranes (Rizzoli and Galzigna 1970). Of course there is the possibility that GHBA acts on REM sleep

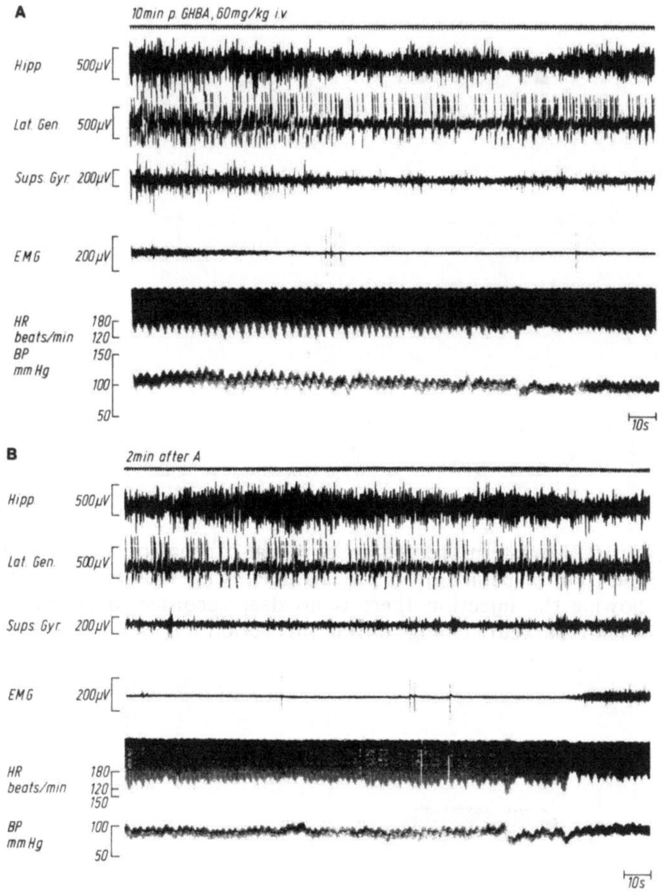

Fig. 9. Transition from GHBA-induced nonREM sleep to REM sleep *(panel A)*. Recordings as in Fig. 1. Apart from the electrophysiologic changes there is an attenuation of respiratory arrhythmia during GHBA-induced REM sleep as is also seen during spontaneous REM sleep. *Panel B* shows the same REM period as *panel A* with the transition to wakefulness

directly via GABA receptors, the changes in DA turnover being a secondary rather than a primary effect. In fact the effects of GHBA on DA mechanisms are comparable to those of GABA (Andén and Stock 1973; Stock et al. 1978); though there are few reports involving GABA in sleep. Jasper et al. (1965) and Jasper and Koyama (1969) found increased release of GABA from the cerebral cortex during a sleep-like stage seen after transecting at the level of the midbrain. Karadzic (1967) and de Mendoza et al. (1973) also obtained evidence relating GABA to REM sleep. They observed a decrease in REM sleep when GABA levels were increased both in cat and rat brains.

On the other hand, if the effects of GHBA on DA mechanisms are causally related to GHBA-induced REM sleep one could explain the failure of low doses of haloperidol to block GHBA-induced REM sleep by assuming that haloperidol might have different effects at different doses.

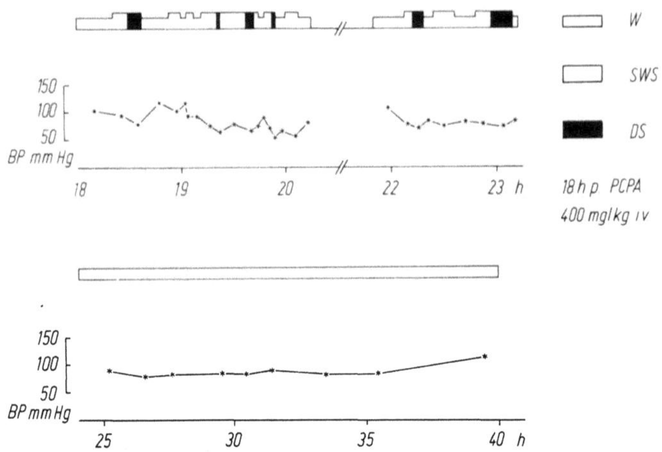

Fig. 10. Long-term recording of a cat's behavior and arterial blood pressure after application of parachlorophenylalanine (*PCPA*, 400 mg/kg i.p.). After a period of 23 h following the injection there is no sleep recorded in the cat. Arterial blood pressure does not markedly change during this period

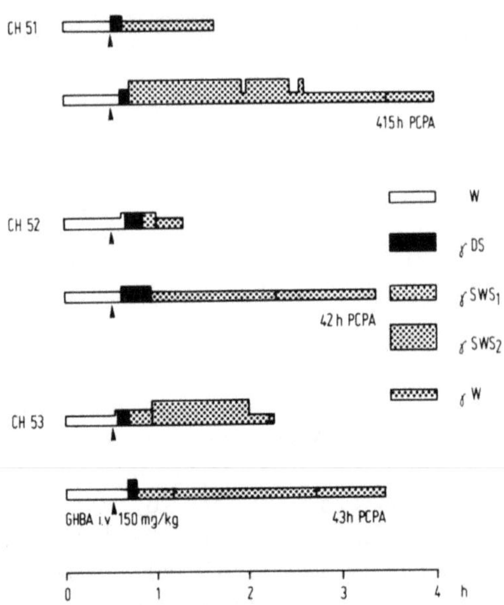

Fig. 11. Diagram as in Fig. 8. Results are shown from experiments in three cats. *GHBA*, 150 mg/kg i.v. induced REM sleep both before and approximately 40 h after application of *PCPA*, 400 mg/kg. The time of *GHBA* injection is indicated by an *arrow*. From each cat two sequences are given: the *upper* showing the effect of *GHBA* alone, the *lower* showing the effect of *GHBA* in the *PCPA*-treated cat

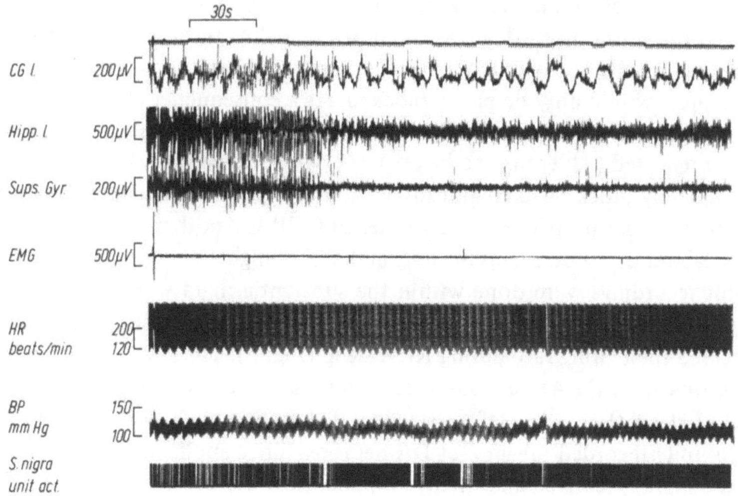

Fig. 12. This figure shows data of the same cat and of the same recording session as in Fig. 5. The transition from GHBA-induced nonREM sleep to REM sleep is given. GHBA, 50 mg/kg has been i.v. injected 5 min before. The same nigral unit is shown as in Fig. 5. It's firing rate is increased in the same way as in Fig. 5 seen during spontaneously occuring REM sleep

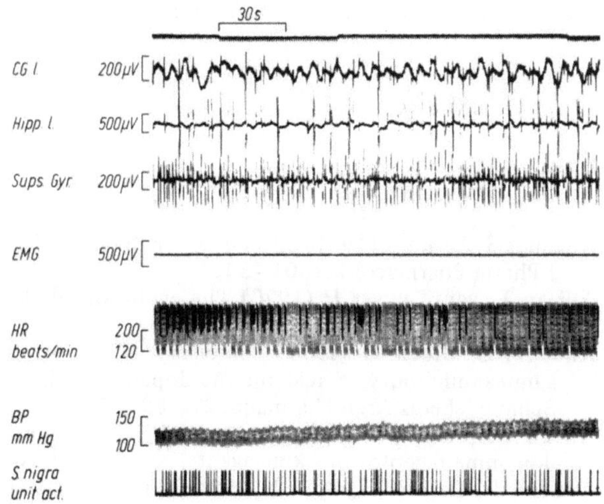

Fig. 13. The same recording session and the same nigral unit is shown as in Figs. 5 and 12. In addition to the low dose of GHBA, 50 mg/kg i.v. the cat received GHBA 200 mg/kg. The figure is taken 23 min after Fig. 12 and 12 min after application of the additional dose of GHBA, 200 mg/kg i.v. The cat is anesthetized. The firing rate of the nigral unit is markedly decelerated

Low doses could yield a paradoxical stimulation of postsynaptically located DA receptors if the net effect of low doses of haloperidol will be a marked blockade of presynaptically located DA autoreceptors, whereas the postsynaptically located DA receptors would only be partly blocked. As a consequence the postsynaptically located DA receptors would be stimulated. In fact such paradoxical effects of haloperidol have been reported (Ahlenius and Engel 1971). In contrast high doses of neuroleptics would completely block postsynaptically located DA receptors and hence counteract the REM sleep-inducing properties of low doses of GHBA. Another finding, obtained recently in this laboratory strengthens the notion that DA might be involved in REM sleep generation. Unit recordings were done within the substantia nigra where the DA fibers originate. They decreased their firing rate during spontaneously occurring nonREM sleep and increased their firing rate during REM sleep (Fig. 5). The same pattern was obtained when the same units (N=4) were observed after i.v. injection of GHBA (50 mg/kg) which induced REM sleep (Fig. 12). Before arriving at definite conclusions, it has to be shown that the units recorded are in fact DA neurons. Since application of higher doses of GHBA (200 mg/kg) reduced and finally stopped the firing of these units there is some evidence that they might be dopaminergic (Fig. 13). Such a response has been shown to be characteristic for nigral DA neurons in rats (Walters et al. 1972).

In conclusion: It is hypothesized that DA acts on REM sleep generation. This is not achieved in an all-or-none fashion. Rather it appears that DA needs to be released and must be present in critical amounts within the synaptic clefts in order to trigger or maintain some – probably phasic – aspects of REM sleep.

Acknowledgments. This work was supported by grants from the Deutsche Forschungsgemeinschaft.

I am greatly indebted to Dr. Mark Knuepfer for his valuable criticism and to Mrs. D. Fischer-Barnicol for her help in preparing the manuscript.

References

Ahlenius S, Engel S (1971) Effects of small doses of haloperidol on timing behaviour. J Pharm Pharmacol 23:301–302

Allison T, van Twyver H (1970) The evolution of sleep. Natural History, New York 79:56–65

Andén NE, Stock G (1973) Inhibitory effect of gammahydroxybutyric acid and gammaaminobutyric acid on the dopamine cells in the substantia nigra. Naunyn-Schmiedebergs Arch Pharmacol 279:89–92

Andén NE, Grabowska-Andén M (1979) Presynaptic and postsynaptic effects of dopamine receptor blocking agents. Adv Neurol 24–235–245

Andén NE, Corrodi H, Fuxe K, Hökfelt T (1965) Mapping out of catecholamine and 5-hydroxytryptamine neurons innervating the telencephalon and diencephalon. Life Sci 4:1275–1281

Andén NE, Magnusson T, Stock G (1973) Effects of drugs influencing monoamine mechanisms on the increase in brain dopamine produced by axotomy or treatment with gammahydroxybutyric acid. Naunyn-Schmiedebergs Arch Pharmacol 278: 363–372

Angrist B, Gershon S (1970) The phenomenology of experimentally induced amphetamine psychosis, preliminary observation. Biol Psychiatry 2:95–107

Aserinsky E (1965) Periodic respiratory pattern occurring in conjunction with eye movements during sleep. Science 150:763–766

Aserinsky E, Kleitman N (1953) Regularly occurring periods of eye motility and concurrent phenomena during sleep. Science 118:273–274

Baldissera F, Broggi G, Mancia M (1966) Monosynaptic and polysynaptic spinal reflexes during physiological sleep and wakefulness. Arch. Ital Biol 104:112–133

Bartholini G (1976) Differential effect of neuroleptic drugs on dopamine turnover in the extrapyramidal and limbic system. J Pharm Pharmacol 28:429–443

Baumgarten HG, Björklund A, Holstein AF, Nobin A (1972) Chemical degeneration of indolamine axons in rat brain by 5-6 HT. An ultrastructural study. Z. Zellforschung 129:256–271

Baust W, Bohnert B (1969) The regulation of heart rate during sleep. Exp Brain Res 7:169–180

Baust W, Weidinger H, Kirchner F (1968) Sympathetic activity during natural sleep and arousal. Arch Ital Biol 106:379–390

Beck U, Schenck E, Ischinger TH (1973) Spinale und bulbäre Reflexe im Schlaf beim Menschen. Arch Psychiatr Nervenkr 217:157–168

Berger RJ (1961) Tonus of extrinsic laryngeal muscles during sleep and dreaming. Science 134:840

Berlucchi G, Munson JB, Rizzolatti G (1967) Changes in click-evoked response in the auditory system and the cerebellum of free moving cats during sleep and waking. Arch Ital Biol 105:118–137

Berman AL (1968) The brainstem of the cat. University of Wisconsin Press, Madison, Milwaukee

Bizzi E, Pompeiano O, Somogti J (1964) Spontaneous activity of single vestibular neurons of unrestrained cats during sleep and wakefulness. Arch Ital Biol 102: 308–330

Björklund A, Lindvall O (1978) The meso-telencephalic dopamine neuron system: a review of its anatomy. In: Livingston KE, Hornykiewicz O (eds) Limbic mechanism. Plenum, New York, pp 307–331

Björklund A, Moore RY, Nobin A, Stenevi U (1973) The organization of tubero-hypophyseal and reticulo-infundibular catecholamine neuron systems in the rat brain. Brain Res 51:171–191

Björklund A, Falck B, Nobin, A, Stenevi U (1974) Organization of the dopamine and noradrenaline innervations of the median eminence-pituitary region in the rat. In: Neurosecretion – The Final Neuroendocrine Pathway, VIth International Symposium on Neurosecretion, London, 1973, Springer, Berlin Heidelberg New York

Björklund A, Lindvall O, Nobin A (1975) Evidence of an incerto-hypothalamic dopamine neurone system in the rat. Brain Res 89:29–42

Bliwise D, Coleman R, Bergmann B, Wincor M, Pivic RT, Rechtschaffen A (1974) Facial muscle tonus during REM and NREM sleep. Psychophysiology 11:497–508

Bonasegla F, Menegati E (1968) Effect of the administration of L-Dopa on the phases of sleep. Rass Studi psichiat 57:797–804

Bristow JD, Hanour AJ, Pickering TG, Sleight P (1969) Cardiovascular and respiratory changes during sleep in normal and hypertensive subjects. Cardiovasc Res 4:476–485

Brodal A (1957) The reticular formation of the brain stem. Anatomical aspects and functional correlations. Oliver & Boyd, Edinburgh

Brooks DC (1969) Localization and characteristics of the cortical waves associated with eye movement in the cat. Exp Neurol 22:603–613

Buguet A (1969) Monoamines et sommeil. In: Etude des relations entre les structures monoaminergiques du pont et les pointes pontogeniculo-occipitales du sommeil. Thèse Médecine, Lyon

Buguet A, Petitjèan F, Jouvet M (1970) Suppression des pointes PGO du sommeil par lésion ou injection in situ de 6-hydroxydopamine au niveau du tegmentum pontique. C R Soc Biol (Paris) 164:2293–2298
Bülow K (1963) Respiration and wakefulness in man. Acta Physiol Scand [Suppl] 59: 209
Carlsson A, Lindquist M (1963) Effects of chlorpromazine or haloperidol on formation of 3-methoxytyramine and normetanephrine in mouse brain. Acta Pharmacol Toxicol (Copenh.) 20:140–144
Carlsson A, Kehr W, Lindquist M, Magnusson T, Atack CV (1972) Regulation of monoamine metabolism in the central nervous system. Pharmacol Res 24:371–384
Castaldo V, Krynicki VE, Crade M (1973) L-Dopa and Rem sleep in normal and mentally retarded subjects. Biol Psychiatry 6:295–299
Candia O, Favale E, Guissari A, Rossi GF (1962) Blood pressure during natural sleep and during sleep induced by electrical stimulation of the brain stem reticular formation. Arch Ital Biol 100:216–233
Cespuglio R, Riou F, Faradji H, Ponchon JT, Buda M, Gonon F, Jouvet M, Pujol JF (1979) The release of monoamines during the sleep waking cycle as studied with the voltammetric technique. Neuroscience Lett [Suppl] 3:323
Chase MH (1978) State-dependent reversal of a brainstem reflex in Felix domesticus. In: Ferrendelli JA (ed) Aspects of behavioral neurobiology, vol 3. Society for Neurosciene Bethesda, Maryland, pp 33–65
Chase MH, McGinty DJ, Sterman MB (1968) Cyclic variations in the amplitude of a brain stem reflex during sleep and wakefulness. Experientia 24:47–48
Cheramy A, Nieoullon A, Glowinski J (1977) Stimulating effects of hydroxybutyrate on dopamine release from the caudate nucleus and the substantia nigra of the cat. J Pharmacol Exp Ther 203:283–293
Chu NS, Bloom FE (1973) Norepinephrine-containing neurons; changes in spontaneous discharge patterns during sleeping and waking. Science 179:908–910
Chu NS, Bloom FE (1974) The catecholamine-containing neurons in the cat dorsolateral pontine tegmentum: distribution of the cell bodies and some axonal projections. Brain Res 66:1–21
Cianchetti C, Masala C, Mangoni A, Gessa GL (1980) Suppression of REM and delta sleep by amorphine in man: A dopamine mimetic effect. Psychopharmacology 67:61–65
Coccagna G, Montovani M, Brignani F, Manrino A, Lugaresi E (1971) Arterial pressure changes during spontaneous sleep in man. Electroencephalogr Clin Neurophysiol 31:277–281
Conrad LCA, Leonard CM, Pfaff DW (1974) Connections of the median and dorsal raphe nuclei in the rat: an autoradiograpic and degeneration study. J Comp Neurol 156:179–206
Crow TJ, Johnstone EC, Longden A, Owen F (1978) Dopamine and schizophrenia. Adv Biochem Psychopharmacol 9:301–309
Crow TJ (1980) Molecular pathology of schizophrenia: More than one disease process? Br Med J 280:66–68
Dahlström A, Fuxe K (1964) Evidence for the existence of monoamine-containing neurons in the central nervous system. I. Demonstration of monoamines in the cell bodies of brain stem neurons. Acta Physiol Scand 62:1–55
Dahlström A, Fuxe K (1965) Evidence for the existence of monoamine neurons in the central nervous system. II. Experimentally induced changes in the intraneuronal amine levels of bulbospinal neuron systems. Acta Physiol Scand [Suppl] 64:247
De Mendoza JL, Guthier J, Rodi P, Gottesman C, Roux R (1973) Influence de l'élevation du taux de l'acide gamma aminobutyrique dans le système nerveux central sur les différentes phases du cycle veille-sommeil chez le rat. C R Soc Biol (Paris) 167:73–79

Dement W (1958) The occurrence of low voltage, fast, electroencephalogram patterns during behavioral sleep in the cat. Electroencephalogr Clin Neurophysiol 10: 291–296

Dement WC, Kleitman N (1957) Cyclic variations in EEG during sleep and their relation to eye movements, body mobility and dreaming. Electroencephalogr Clin Neurophysiol 9:673–690

Doherty JD, Hattox SE, Carter Snead O, Roth RH (1978) Identification of endogenous γ-hyroxybutyrate in human and bovine brain and its regional distribution in human, guinea pig and rhesus monkey brain. J Pharmacol Exp Ther 207:103–139

Duron B (1969) Activité électrique spontanée des muscles intercostaux et le diaphragme chez l'animal chronique. J Physiol (Paris) 61 [Suppl 2]:282–283

Dusan-Peyrethon D, Froment JL (1968) Effets du disulfiram sur les états de sommeil chez le chat. C R Soc Biol (Paris) 162:2141–2145

Dusan-Peyrethon D, Peyrethon J, Jouvet M (1968) Suppression élective du sommeil paradoxal chez le chat par alpha méthyl-dopa. C R Soc Biol (Paris) 162:116–118

Evarts EV (1962) Activity of neurons in visual cortex of the cat during sleep with low voltage fase EEG activity. J Neurophysiol 25:812–816

Evarts EV (1964) Temporal patterns of discharge of pyramidal tract neurons during sleep and waking in the monkey. J Neurophysiol 27:152–171

Fibiger HC, Phillips AG (1979) Dopamine and the neural mechanisms of reinforcement. In: Horn AS, Korf J, Westerink BHC (eds) The neurobiology of dopamine. Academic Press, London New York San Francisco, pp 597–615

Findlay ALR, Hayward JN (1969) Spontaneous activity of single neurones in the hypothalamus of rabbits during sleep and waking. J Physiol (Lond) 201:237–258

Fisher C, Gross J, Zunch J (1965) Cycle of penile erection synchronous with dreaming (REM) sleep. Arch Gen Psychiatry 12:29–45

Fuxe K (1964) Cellular localization of monoamines in the median eminence and infundibular system of some mammals. Z Zellforsch 61:710–724

Fuxe K (1965) Evidence for the existence of monoamine neurons in the central nervous system. IV Distribution of monoamine nerve terminals in the central nervous system. Acta Physiol Scand 64 [Suppl 247] 39–85

Gadea-Ciria M, Stadler H, Lloyd K, Bartholini G (1973) Acetylcholine release within the cat striatum during the sleep-wakefulness cycle. Nature 243:518–519

Gaillard JM, Monema A (1977) Modification of dream content after preferential blockade of mesolimbic and mesocortical dopaminergic systems. J Psychiatry Res 13:247–256

Gassel M, Pompeiano O (1965) Fusimotor function during sleep in unrestrained cats. Arch Ital Biol 103:347–368

Gassel M, Marchiafava PL, Pompeiano O (1964) Tonic and phasic inhibition of spinal reflexes during deep desynchronized sleep in unrestrained cats. Arch Ital Biol 102: 471–499

Griesinger W (1868) Berliner medizinisch-psychologische Gesellschaft, Sitzung 29. 1. 1867. Arch Psychiatr Nervenkr 1:200–204

Gschwend J (1970) Der Haltetonus der Katze im Schlaf. Brain Res 22:29–45

Guilleminault Ch, Dement WC (1978) Sleep apnea syndromes. Alan R. Liss. Inc., New York

Hartmann E (1976) Schizophrenia: A theory. Psychopharmacology (Berlin) 49:1–15

Hartmann E (1978) Effects of psychotropic drugs on sleep: The catecholamines and sleep. In: Lipton MA, Di Mascio A, Killam KF (eds) Psychopharmacology: A generation of progress. Raven Press, New York, pp 711–728

Hartmann E, Zwilling G (1975) Intraventricular norepinephrine and dopamine: differential effects on sleep in the rat. (Abstract) Neuroscience 1:248

Hartmann E, Bridwell TJ, Schildkraut JJ (1971) Alpha-methylparatyrosine and sleep in the rat. Psychopharmacology (Berlin) 21:157–164

Henley K, Morrison AR (1974) A re-evaluation of the effects of lesions of the pontine tegmentum and locus coeruleus on phenomena of paradoxical sleep in the cat. Acta Neurobiol Exp 34:215–232

Hernandez-Péon R (1962) Sleep induced by localized or chemical stimulation of the forebrain. Electroencephalogr Clin Neurophysiol 14:423–424

Hernandez-Péon R, Chavez-Ibarra G (1963) Sleep induced by electrical or chemical stimulation of the forebrain. Electroencephalogr Clin Neurophysiol [Suppl] 24:188–198

Hobson JA (1980) Toward a cellular neurophysiology of the reticular formation: Conceptual and methodological milestones. In: Hobson JA, Brazier AB (eds) IBRO Monograph series, vol 6. Raven Press, New York, pp 7–29

Hobson JA, McCarley RW (1971) Cortical unit activity in sleep and waking. Electroencephalogr Clin Neurophysiol 30:97–112

Hobson JA, McCarley RW, Pivik RT, Freedman R (1974) Selective firing by cat pontine brain-stem neurons in desynchronized sleep. J Neurophysiol 37:497–511

Hobson JA, McCarley RW, Wyzinski PW (1975) Sleep cycle oscillation: reciprocal discharge by two brain stem neuronal groups. Science 189:55–58

Hodes R, Dement WC (1964) Depression of electrically induced reflexes ("H-reflexes") in man during low voltage EEG sleep. Electroencephalogr Clin Neurophysiol 17: 617–629

Hökfelt T, Fuxe K (1972) On the morphology and the neuroendocrine role of the hypothalamic catecholamine neurons. In: Brain-Endocrine Interaction, Median Eminence Structure and Function. International Symposium. Karger, Basel, pp 181–223

Hornykiewicz O (1979) Brain dopamine in Parkinson's disease and other neurological disturbances. In: Horn AS, Korf J, Westerink BHC (eds) The neurobiology of dopamine. Academic Press, London New York San Francisco, pp 633–654

Hubel DH (1959) Single unit activity in striate cortex of unrestrained cats. J Physiol (Lond) 147:226–238

Jacobs BL, McGinty DJ (1971) Amygdala unit activity during sleep and waking. Exp Neurol 33:1–15

Jacobs BL, Trulson ME (1979) Dreams, hallucinations and psychosis – the serotonin connection TINS 11:276–280

Jasper HH, Koyama J (1969) Rate of release of amino acids from cerebral cortex in the cat as affected by brainstem and thalamic stimulation. Can J Physiol Pharmacol 10:889–905

Jasper HH, Tessier J (1971) Acetylcholine liberation from cerebral cortex during paradoxical (REM) sleep. Science 172:601

Jasper HH, Kahn RT, Elliot KAC (1965) Amino acids released from the cerebral cortex in relation to its state of activation Science 147:1448–1449

Jessell TM, Cuello AC, Iversen LL (1976) Release of dopamine from dendrites in rat substantia nigra. Nature 260:258–260

Jones BE, Bobillier P, Pin C, Jouvet M (1973) The effect of lesions of catecholamine-containing neurons upon monoamine content of the brain and EEG and behavioural waking in the cat. Brain Res 58:157–177

Jones BE, Harper TS, Halaris EA (1977) Effects of locus coeruleus lesions upon cerebral monoamine content, sleep wakefulness states and the response to amphetamine in the cat. Brain Res 124:473–496

Jouvet M (1959) Corrélations électromyographiques du sommeil chez le chat décortiqué et mésencéphalique chronique. C R Biol 153:422–425

Jouvet M (1962) Recherches sur les structures nerveuses et les mécanismes responsables des différentes phases du sommeil physiologique. Arch Ital Biol 100:125–206

Jouvet M (1969) Biogenic amines and the states of sleep. Science 163:32–41

Jouvet M, Michel F, Mounier D (1960) Analyse électroencéphalographique comparée du sommeil physiologique chez le chat et chez l'homme. Rev Neurol 103:189–205

Jouvet M (1972) The role of monoamines and acetylcholine-containing neurons in the regulation of the sleep-waking cycle. Erg Physiol Biochem Pharmacol 64: 166–307

Jouvet M (1973a) A possible role of the catecholamine – containing neurons of the brain stem of the cat in the sleep-waking cycle. Acta Physiol (Pol) 24:5–19

Jouvet M (1973b) Essai sur le rêve. Arch Ital Biol 3:564–576

Jouvet M (1977) Neuropharmacology of the sleep-waking cycle. In: Iversen LL, Iversen SD, Snyder SH (eds) Handbook of psychopharmacology, vol 8. Plenum Press, New York London, pp 233–292

Jouvet M (1978) Does a genetic programming of the brain occur during paradoxical sleep? In: Buser PA, Rougeul-Buser A (eds) Cerebral correlates of conscious experience. Inserm Symposium No. 6, Elsevier North Holland, Biomedical Press, pp 245–261

Jouvet M (1979) What does the cat dream about? TINS 11:280–282

Jouvet M, Michel F (1959) Correlations électromyographiques du sommeil chez le chat décortiqué et mésencéphalique chronique. C R Soc Biol 153:422–425

Jouvet M, Mounier D (1960) Analyse électroencéphalographique comparée du sommeil physiologique chez le chat et chez l'homme. Rev Neurol 103:189–205

Jouvet M, Jeannerod M, Delorme F (1965) Organisation du système responsable de l'activité phasique au cours du sommeil paradoxal. C R Soc Biol 7:1599–1604

Jovanović UJ (1971) Erektionen im Schlaf und vegetative Funktionen. J Neuro-Visceral Relations 32:200–216

Kafi S, Bouras C, Constantinidis J, Gaillard JM (1977) Paradoxical sleep and brain catecholamines in the rat after single and repeated administration of α-methyl-paratyrosine. Brain Res 135:123–134

Kafi S, Gaillard JM (1976) Brain dopamine receptors and sleep in the rat: effects of stimulation and blockade. E J Pharmacol 38:357–363

Kamberi JA, Mical RS, Porter JC (1971) Effect of anterior pituitary perfusion and intraventricular injection of catecholamines on prolactin release. Endocrinology 88:1012–1020

Karacan J, Goodenough DR, Shapiro A, Starker S (1966) Erection cycle during sleep in relation to dream anxiety. Arch Gen Psychiatry 15:183–189

Karadzic V (1967) Effect of raised levels of gamma-aminobutyric acid in the central nervous system on sleep phases in the cat. Acta Med Jugosl 20:37–44

Kay DC, Blackburn AB, Buckingham JA, Karacan J (1976) Human pharmacology of sleep. In: Williams RL, Karacan J (eds) Pharmacology of sleep. Wiley, New York London Sydney Toronto, pp 83–210

Kehr W, Carlsson A, Lindqvist M (1972) A method for the determination of 3,4-dihydroxyphenylalanine (DOPA) in brain. Naunyn-Schmiedebergs Arch Pharmacol 274:273–280

Kendel K, Beck U, Wita C, Hohneck E, Zimmermann H (1972) Der Einfluß von L-Dopa auf den Nachtschlaf bei Patienten mit Parkinson-Syndrom. Arch Psychiatr Nervenkr 216:82–100

Khatri JM, Freis ED (1967) Hemodynamic changes during sleep. J Appl Physiol 22: 867–873

King CD (1971) The pharmacology of rapid eye movement sleep. Adv Pharmacol Chemotherap 9:1–91

King CD, Jewett EJ (1971) The effects of α-methyltyrosine on sleep and brain norepinephrine in cats. J Pharmacol Exp Ther 177:188–195

Kirchner F (1973) Correlations between changes of activity of the renal sympathetic nerve and behavioural events in unrestrained cats. Basic Res Card 69:243–256

Kiyono S, Jeannerod M (1967) Relations entre l'activité géniculée phasique et les mouvements oculaires chez le chat normal et sous reserpine. C R Soc Biol (Paris) 161:1607–1611

Kleinberg DL, Noel GL, Frantz AG (1971) Chlorpromazine stimulation and l-dopa suppression of plasma prolactin in man. J Clin Endocrinol Metab 33:873–876

Koella WP, Feldstein A, Czicman JS (1968) The effect of parachlorophenylalanine on the sleep of cats. Electroencephalogr Clin Neurophysiol 25:481–490

Korf J, Zielman M, Westerink BHC (1976) Dopamine release in substantia nigra? Nature 260:257–258

Kovacevic R, Radulovacki M (1976) Monoamine changes in the brain of cats during slow-wave sleep. Science 193:1025–1027

Kubota K, Tanaka R (1968) Fusimotor unit activities and natural sleep in the cat. Jpn J Physiol 18:43–58

Laborit H, Jouany JM, Gerard J, Fabiani F (1960) Généralités concernant l'étude expérimentale et l'emploi clinique du γ-hydroxybutyrate de Na. Aggressologie 1: 397–406

Laffont F, Autret A, Minz M, Beillevaire T, Cathala HP, Castaigne P (1978) Sleep respiratory arrhythmias in control subjects, narcoleptics and non-cataleptic hypersomniacs. Electroencephalogr Clin Neurophysiol 44:697–705

Lamarre Y, Filion M, Cordeau JP (1971) Neuronal discharge of the ventrolateral nucleus of the thalamus during sleep and wakefulness in the cat-I.Spontaneous activity. Exp Brain Res 12:48–498

Leichnetz GR (1972) Relationship of spontaneous vagal activity to wakefulness and sleep in the cat. Exp Neurol 35:194–210

Lindvall O, Björklund A (1978) Organization of catecholamine neurons in the rat central nervous system. In: Iversen LL, Iversen SD, Snyder SH (eds) Handbook of psychopharmacology, vol 9. Plenum, New York, pp 139–231

Mancia G, Giorgio B, Adams OB, Zanchetti A (1970) Vasomotor regulation during sleep in the cat. Am J Physiol 220:1086–1093

Matsumoto J, Jouvet M (1964) Effects de la réserpine, Dopa et 5-HTP sur les deux états de sommeil. C R Soc Biol 158:2137–2140

Matsuzaki M, Tagaki H (1967) Sleep induced by sodium butyrate in the cat. Brain Res 4:206–222

McCarley RW (1978) Control of sleep-waking state alteration in felix domesticus. In: Ferrendelli JA (ed) Aspects of behavioral neurobiology, vol 3. Society for Neuroscience, Bethesda, Maryland, pp 90–128

McCarley RW (1980) Mechanisms and models of behavioral state control. In: Hobson JA, Brazier MAB (eds) The reticular formation revisited. IBRO Monogr Series, vol 6. Raven Press, New York, pp 375–403

McCarley RW, Hobson JA (1971) Single neurons activity in cat gigantocellular tegmental field: selectivity of discharge in desynchronized sleep. Science 174: 1250–1252

McCarley RW, Hobson JA (1975) Discharge patterns of cat pontine brain stem neurons during desynchronized sleep. J Neurophysiol 38:751–766

McGinty DJ (1973) Neurochemically defined neurons: behavioral correlates of unit activity of serotonin-containing neurons. In: Phillips MI (ed) Brain unit activity during behavior. Thomas, Springfield, Ill., pp 244–276

McGinty DJ, Harper RM (1976) Dorsal raphe neurons: depression of firing during sleep in cats. Brain Res 101:569–575

McWilliams JA (1923) Blood pressure and heart action in sleep and dreams. Br Med J 2:1196–1200

Mereu GP, Scarnati E, Paglietti E, Pellegrini O, Quarantotti B, Chessa P, Di Chiara G, Gessa GL (1979) Sleep induced by low doses of apomorphine in rats. Electroencephalogr Clin Neurophysiol 46:214–219

Mikiten T, Niebyl P, Hendley C (1961) EEG desynchronization during behavioral sleep associated with spike discharges from the thalamus of the cat. Fed Proc 20:327

Millar-Craig MW, Bishop CN, Raftery EB (1978) Circadian variation of blood-pressure. Lancet I/2:795–797

Mink WD, Best PJ, Olds J (1967) Neurons in paradoxical sleep and motivated behavior. Science 158:1335–1337

Monti JM (1968) The effect of haloperidol on the sleep cycle of the cat. Experientia 24:1143–1144

Morales F, Chase M (1978) Intracellular recording of lumbar motoneuron membrane potential during sleep and wakefulness. Exp Neurol 62:821–827

Morales F, Chase M (1980) Synaptic actions upon lumbar alpha motoneurons during active sleep. Sleep Res 8:32

Morgane PJ, Stern WC (1974) Chemical anatomy of brain circuits in relation to sleep and wakefulness. In: Weitzmann ED (ed) Advances in sleep research, vol 1. Spectrum, New York, pp 1–131

Moruzzi G (1963) Active processes in the brain stem during sleep. Harvey Lect 58: 233–297

Moruzzi G (1972) The sleep-waking cycle. Rev Physiol Biochem Pharmacol 46:1–165

Mouret J (1975) Differences in sleep in patients with parkinson's disease. Electroencephalogr Clin Neurophysiol 38:653–657

Mouret J, Jeannerod M, Jouvet M (1963) L'activité électrique du systéme visuel au cours de la phase paradoxale du sommeil chez le chat. J Physiol (Paris) 55:305–306

Mouret J, Renaud B, Blois R (1978) Central dopamine metabolism in human paradoxical sleep disorders. In: Usdin E, Kopin IJ, Barchas J (eds) Catecholamines: basic and clinical frontiers, vol 2. Pergamon Press, New York Oxford Toronto Sydney Frankfurt Paris, pp 1581–1583

Mukhametov LM, Rizzolatti G, Tradardi V (1970) Spontaneous activity of neurons of nucleus reticularis thalami in freely moving cats. J Physiol (Lond) 210:651–667

Murphy EH, Starr A (1971) Evoked responses to electrical stimulation of the auditory pathways during the wake sleep cycle. Brain Res 35:491–500

Nakamura Y, Goldberg LJ, Chandler SH, Chase MH (1978) Intracellular analysis of trigeminal motoneuron activity during sleep in the cat. Science 199:204–207

Nauta WJH, Kuypers HGJM (1958) Some ascending pathways in the brain stem reticular formation. In: Jasper HH (ed) Reticular formation of the brain. Little & Brown, Boston, pp 3–30

Nelson JP, McCarley WR, Hobson JA (1980) PGO burst neurons discharge with high specificity and coherence to primary waves in the ipsilateral LGB. Sleep Res 8:34

Netick A, Orem J, Dement W (1977) Neuronal activity specific to REM sleep and its relationship to breathing. Brain Res 120:197–207

Nobin A, Björklund A (1973) Topography of the monoamine neuron system in the human brain as revealed in fetuses. Acta Physiol Scand [Suppl] 388:1–40

Noda H, Adey WR (1970 a) Firing of neuron pairs in cat association cortex during sleep and wakefulness. J Neurophysiol 33:672–684

Noda H, Adey WR (1970 b) Firing variability in cat association cortex during sleep and wakefulness. Brain Res 18:513–526

Noda H, Adey WR (1970 c) Changes in neuronal activity in association cortex of the cat in relation to sleep and wakefulness. Brain Res 19:263–275

Noda H, Manohar S, Adey WR (1969) Spontaneous activity of cat hippocampal neurons in sleep and wakefulness. Exp Neurol 24:217–231

Ohlmeyer P, Brilmayer H (1947) Periodische Vorgänge im Schlaf. Pfluegers Arch 2: 249–250

O'Keefe J, Bouma H (1969) Complex sensory properties of certain amygdala units in the freely moving cat. Exp Neurol 23:384–398

Olianas MC, De Montis MG, Mulas G, Tagliamonte A (1976) Synergistic potentiation of haloperidol-induced catalepsy by γ-butyrolactone. J Neurochemistry 27:821–823

Oomura Y, Ooyama H, Naka F, Yamamoto T, Ono T, Kobayashi N (1969) Some stochastical patterns of single unit discharges in the cat hypothalamus under chronic conditions. Ann NY Acad Sci 157:666–689

Parker DC, Rossmann LG, Van der Laan EF (1974) Relation of sleep-entrained human prolactin release to REM-Non REM cycles. J Clin Endocrinol Metab 38:646–651

Parmeggiani PL, Sabattini L (1972) Electromyographic aspects of postural, respiratory and thermoregulatory mechanisms in sleeping cats. Electroencephalogr Clin Neurophysiol 33:1–13

Perret JL, Tapissier J, Jouvet M (1979) Insomnie et mémoire. A propos d'une observation de dégénérescence striato-nigrique. Electroencephalogr Clin Neurophysiol 47: 499–502

Petitjean F, Jouvet M (1970) Hypersomnie et augmentation de l'acide 5-hydroxy-indolacétique cérébal par lésion isthmique chez le chat. C R Soc Biol (Paris) 40: 688–693

Phillipson EA, Sullivan CE (1978) Respiratory control mechanism during NREM and REM sleep. In: Guilleminault Ch, Dement WC (eds) Sleep apnea syndromes. Allan R Liss, New York, pp 47:64

Phillipson EA, Murphy E, Kozar LF (1976) Regulation of respiration in sleeping dogs. J Appl Physiol 40:688–693

Pivik RT, McCarley RW, Hobson JA (1977) Eye movement-associated discharge in brain stem neurons during desynchronized sleep. Brain Res 121:59–76

Pompeiano O (1967) The neurophysiological mechanisms of the postural and motor events during desynchronized sleep. Res Publ Assoc Res Nerv Ment Dis 45:351–423

Pompeiano O (1970) Sleep mechanisms. In: Jasper MM (ed) Basic mechanisms of the epilepsies. Little Brown, Boston, pp 453–473

Pompeiano O (1980) Cholinergic activation of reticular and vestibular mechanisms controlling posture and eye movements. In: Hobson JA, Brazier MAB (eds) The reticular formation revisited. IBRO Monograph Series, vol 6. Raven Press, New York, pp 473–512

Pompeiano O, Hoshino K (1976) Tonic inhibition of dorsal pontine neurons during the postural atonia produced by an anticholinesterase in the decerebrate cat. Arch Ital Biol 114:310–340

Praag van HM (1979) Dopamine and the development of disorders of human behavior. In: Horn AS, Korf J, Westerink BHC (eds) The neurobiology of dopamine, vol 36. Academic Press, London, pp 655–677

Pribram KH, McGuinnes D (1975) Arousal, activation and effort in the control of attention. Psychol Rev 82:567–574

Pujol JF, Buguet A, Froment JL, Jones B, Jouvet M (1971) The central metabolism of serotonin in the cat during insomnia: neurophysiological and biochemical study after p-chlorophenylalanine or destruction of the raphe system. Brain Res 29: 195–212

Putkonen PTS (1979) Mechanisms in the control of sleep stages. In: Priest RG, Pletscher A, Ward J (eds) Sleep research proceedings of the northern european symposium on sleep research. MTP Press Ltd., Lancaster, England, pp 19–34

Randrup A, Munkvad J (1968) Behavioral stereotypies induced by pharmacologic agents. Pharmacopsychiatr Neuropsychopharmacol 1:18–26

Reis DJ, Moorhead D, Wooten GF (1969) Differential regulation of blood flow to red and white muscle in sleep and defense behavior. Am J Physiol 217:541–546

Remmers JE, Bartlett D, Putnam MD (1976) Changes in the respiratory cycle associated with sleep. Respir Physiol 28:227–238

Rizzoli AA, Galzigna L (1970) Molecular mechanism of unconscious state induced by butyrate. Biochem Pharmacol 19:2727–2736

Roffwarg HP, Muzio JN, Dement WC (1966) Ontogenetic development of the human sleep-dream cycle. Science 152:604–619

Roth RH, Suhr Y (1970) Mechanism in brain dopamine and its relationship to "sleep". Biochem Pharmacol 19:3001–3012

Roussel B (1967) Monoamines et sommeils: suppression du sommeil paradoxal et diminution de la noradrénaline cérébrale par des lésions des noyaux LC. Thése de Médicine, Tixier, Lyon

Roussel B, Buguet A, Bobillier P, Jouvet M (1967) Locus coeruleus-sommeil paradoxal et noradrénaline cérébrale. C R Soc Biol (Paris) 161:2537–2541

Sakai K (1980) Some anatomical and physiological properties of pontomesencephalic tegmental neurons with special reference to the PGO waves and postural atonia during paradoxical sleep in the cat. In: Hobson JA, Brazier MAB (eds) The reticular formation revisited. IBRO Monograph Series, vol 6. Raven Press, New York, pp 427–447

Sakakura H (1968) Spontaneous and evoked unitary activities of cat lateral geniculate neurons in sleep and wakefulness. Jpn J Physiol 18:23–42

Sakakura H, Iwama K (1967) Effects of bilateral eye enucleation upon single unit activity of the lateral geniculate body in free behaving cats. Brain Res 6:667–678

Sastre JP (1978) Effets des lesions du tegmentum pontique sur l'organisation des etats de sommeil chez le chat. Thèse, Lyon

Scheibel AB, Scheibel ME, Mollica A, Moruzzi G (1955) Convergence and interaction of afferent impulses on single units of reticular formation. J Neurophysiol 18: 309–331

Sheu Y Sh, Nelson JP, Bloom FE (1974) Discharge patterns of cat raphe neurons during sleep and waking. Brain Res 73:263–276

Siegel JM, McGinty DJ, Breedlove SM (1977) Sleep and waking activity of pontine gigantocellular field neurons. Exp Neurol 56:553–573

Siegel JM (1979) Behavioral functions of the reticular formation. Brain Res Rev 1: 69–105

Sitaram N, Gillin JC (1979) Acetylcholine: possible involvement in sleep and analgesia. In: Davis KL, Berger PA (eds) Brain acetylcholine and neuropsychiatric disease. Plenum Press, New York, pp 311:343

Smyth HS, Sleight P, Pickering GW (1969) The reflex regulation of arterial pressure during sleep in man; a quantitative method of assessing baroreflex sensitivity. Circulation Res 24:109–121

Snyder F, Hobson AJ, Morrison DF, Goldfrank F (1964) Changes in respiration, heart rate, and systolic blood pressure in human sleep. J Appl Physiol 19:417–422

Steriade M, Hobson JA (1976) Neuronal activity during the sleep-waking cycle. Prog Neurobiol 6:155–376

Stock G, Magnusson T, Andén NE (1973) Increase in brain dopamine after axotomy or treatment with gammahydroxybutyric acid due to elimination of the nerve impulse flow. Naunyn-Schmiedebergs Arch Pharmacol 278:347–361

Stock G, Heidt H, Buss J, Schlör KH (1978) Sleep patterns in cat induced by gammahydroxybutyric acid. Electroencephalogr Clin Neurophysiol 44:523–527

Stock G, Steiger A, Heidt H (1980a) Gammahydroxybutyrate-induced sleep and anaesthesia can be differentiated by autonomic parameters. Sleep Res 8:70

Stock G, Steinbrenner J, Kummer P (1980b) Supersensitivity of dopamine-autoreceptors. The effect of gammabutyrolactone in long-term haloperidol treated rats. J Neural Transm 47:145–151

Swanson JLW, Hartmann BK (1975) The central adrenergic system: an immunofluoresence study of the location of cell bodies and their afferent connections in the rat utilizing dopamine-β-hydroxylase as a marker. J Comp Neurol 163:467–507

Takahashi Y (1979) Growth hormone secretion related to the sleep and waking rhythm. In: Drucker-Colin R, Shkurovich M, Sterman MB (eds) The functions of sleep. Academic Press, New York, pp 113–145

Trulson ME, Jacobs BL (1978) Raphe unit activity in freely moving cats: Correlation with the level of behavioral arousal. Brain Res 163:135–150

Trulson ME, Jacobs BL (1979) Dissociations between the effects of LSD on behavior and raphe unit activity in freely moving cats. Science 205:515–518

Tusiewicz K, Moldofsky H, Bryan AC (1976) Mechanics of the ribcage and diaphragm during sleep. Fed Proc 35:396

Ungerstedt U (1968) 6-Hydroxydopamine induced degeneration of central monoamine neurons. Eur J Pharmacol 5:107–111

Ungerstedt U (1979) Central dopamine mechanisms and unconditioned behavior. In: Horn AS, Korf J, Westerink BHC (eds) The neurobiology of dopamine. Academic Press, London New York San Fransisco, pp 577:596

Ursin R (1972) Differential effect of parachlorophenylalanine on the two slow wave sleep stages in the cat. Acta Physiol Scand 86:278–285

Van Woert MH (1979) The role of cholinergic and dopaminergic interactions in diseases of the central nervous system. In: Davis KL, Berger PA (eds) Acetylcholine and neuropsychiatric disease. Plenum, New York, pp 373–393

Vertes RP (1977) Selective firing of rat pontine gigantocellular neurons during movement and REM sleep. Brain Res 128:146–152

Walters JR, Aghajanian GK, Roth RH (1972) Dopaminergic neurons: Inhibition of firing by γ-hydroxybutyrate. Proc Fifth Congr Pharmacol, vol 600:100

Walters JR, Aghajanian GK, Roth RH (1972) Dopaminergic neurons: Inhibition of chemical and histochemical effects of gammahydroxybutyrate and acute lesions of the nigro-neostriatal pathway. J Pharmacol 186:630–659

Webb P (1974) Periodic breathing during sleep. J Appl Physiol 37:899–903

Wolfson LJ, Sakurada O, Sokoloff L (1977) Effects of γ-butyrolactone on local cerebal glucose utilization in the rat. J Neurochem 29:777–783

Worms P, Lloyd KG (1980) Biphasic effects of direct, but not indirect, GABA mimetics and antagonists on haloperidol-induced catalepsy. Naunyn-Schmiederbergs Arch Pharmacol 311:179–184

Wyatt RJ, Chase TN, Kupfer DJ, Scott J, Snyder F, Sjoerdsma A, Engelman V (1971a) Brain catecholamines and human sleep. Nature 233:63–65

Wyatt RJ, Chase TN, Scott J, Snyder F (1971b) Effect of L-Dopa on sleep of man. Nature 233:63–65

Wyatt RJ, Gillin JC (1976) Biochemistry and human sleep. In: Williams RC, Karacan I (eds) Pharmacology of sleep. Wiley, New York London Sydney Toronto, pp 239–274

Wyatt RJ, Neff NH, Vaughan T, Franz J, Ommaya A (1974) Ventricular fluid 5-Hydroxyindolacetic acid concentrations during human sleep. In: Costa E, Greengard P (eds) Advances in biochemical psychopharmacology, vol 11. Raven Press, New York, pp 193:197

Endocrine and Peptide Functions in the Sleep-Waking Cycle*

RENÉ DRUCKER-COLÍN[1] and CARLOS VALVERDE-R.[2]

Contents

1 Introduction

Ever since the landmark discovery of Aserinsky and Kleitman (1953) that certain periods of sleep were accompanied by conjugate rapid eye movements (REM), which made it possible to distinguish between two phases of sleep, physiologists have been concerned with finding their neurohumoral counterparts. For many years this aspect was dominated by the monoamine theory of sleep, largely due to the very elegant series of experiments carried out by Jouvet and his group (1972). In recent years, partly due to some conflicting evidence bearing on the monoamine theory (Drucker-Colín and Spanis 1976; Gillin et al. 1978; Ramm 1979) and partly due to the enormous interest generated in the neurosciences by hormones and peptides, this theory has fallen slightly into disfavor. This state of affairs should not be taken to signify that

* This work was partially supported by Grant number PCCBNAL 790215 of the Consejo Nacional de Ciencia y Tecnología

1 Depto. de Neurociencias, Centro de Investigaciones en Fisiología Celular, Universidad Nacional Autónoma de México, México 20, D.F.

2 Depto. de Medicina Nuclear y Clinica de Tiroides, Instituto Nacional de Nutrición, México 22, D.F.

the monoamines do not play a role in sleep, but rather that their role should no longer be conceived as being of central importance, since in all probability sleep depends on a number of systems.

Nowadays it is probably impossible to determine all the systems involved in the regulation of the sleep-wake cycle. Nonetheless, in the past decade hormones and peptides have provided sleep research with new perspectives for studying such regulation, due to the fact that the functions of nervous and endocrine cells are closely interrelated. The very core of this functional reciprocity, the neurosecretory capability, constitutes the corner stone of modern neurobiology and of the so-called diffuse neuroendocrine system (DNES). As reviewed elsewhere (Scharrer 1978; Leibson 1979; Pérez-E. and Valverde-R. to be published), neurosecretion is a clearly definable morphofunctional property (Table 1) of a special type of cell that constitutes the crucial final link or common pathway between neural and endocrine regulatory processes. Recent embryogenic studies on the fate of migrating neuroblastic cells have shown that neuroendocrinologically programmed cells are seeded throughout many tissues (Le Douarin 1974; Le Douarin et al. 1980; Pearse 1976, 1977; Pearse and Takor 1976). These findings and those provided by some phylogenetic studies (King 1979; Jackson 1978, 1980) add further support to the DNES concept and suggest that neurosecretory and several endocrine cells derive from a common ancestor. Furthermore, they substantiate the concept that neurosecretion represents the oldest phylogenetic antecedent of intercellular information transference among living animals, as well as the basic common functional property of the two major integrative subsystems: the nervous and the endocrine.

Therefore, under this expanded spectrum of neurohumoral mediation, the conventional boundaries between the classic neurotransmitters and the vast array of neuropeptides and hormones synthesized by cells of the DNES become blurred and artificial. The DNES is envisaged as a single and highly specialized information-carrying

---

The following abbreviations appear in the text:

A: adrenaline; Ach: acetylcholine; ACTH: corticotropin; ANG-II: angiotensin; ARC: arcuate nucleus; AVT: arginine vasotocin; BE-SRD: brain extract from sleep-deprived rats; CA: catecholaminergic; CSF: cerebrospinal fluid; DA: dopamine; DHT: dihydrotestosterone; DNES: diffuse neuroendocrine system; DSIP: delta sleep-inducing peptide; E_2: estradiol; β-END: β-endorphin; ENK: enkephalins; FSH: follicle stimulating hormone; GABA: γ-aminobutyric acid; GAS-CCK: gastrin-cholecystokinin; GH: growth hormone; 5-HT: 5-hydrotryptamine; GNRH or LHRH: gonodotropin-releasing hormone; HHH: hypothalamo hypophysiotropic hormone; Kd: constant of dissociation; LC: locus coeruleus; LH: luteinizing hormone; β-LPH: β-lipotropin; ME: median eminence; MRF: midbrain reticular formation; MSH: releasing factor; α-MSH: melanocyte-stimulating hormone; MUA: multiple unit activity; NA: noradrenaline; NE: norepinephrine; NP: neurophysin; OVLT: organum vasculosum of the lamina terminalis; OX: oxytocin; OXNp: oxytocin-neurophysin; PGO: ponto-geniculo-occipital; PRL: prolactin; PIF: prolactin release-inhibiting factor; PVN: paraventricular nuclei; REM: rapid eye movement; RM: receptor molecules; RTH: retinohypothalmic; SCN: suprachiasmatic nuclei; SFO: subfornical organ; S-HT: serotonin; SON: supraoptic nuclei; SOPV: supraopticoparaventricular; SRIH: somatostatin; S-W: sleep-wakefulness; SWS: slow wave sleep; T_A: thyroxine; T_3: triyodotironine; TIDA: tuberinfundibular dopaminergic system; TRH: thyroliberin; TSH: thyroid stimulating hormone; VMN: ventromedial nucleus; VP: vasopressin; VPN_p: vasopressin-neurophysin

Table 1. Some morphofunctional features of DNES cells

1.	*Distribution, structural, and functional characteristics*
	Universal phyletic distribution
	Widespread location within and outside the central nervous system
	Common neuroectoblastic ancestry
	Excitability
	Intracytoplasmic secretory granules (100–500 nm)
	Neurocrine (paracrine), neuroendocrine, and endocrine functions
2.	*Biochemical characteristics*
	APUD properties
	Masked metachromasia and argyrophilia
	High content of cholinesterase, nonspecific esterases,
	Mitochondrial α-glycerophosphate dehydrogenase and
	neuron-specific enolase

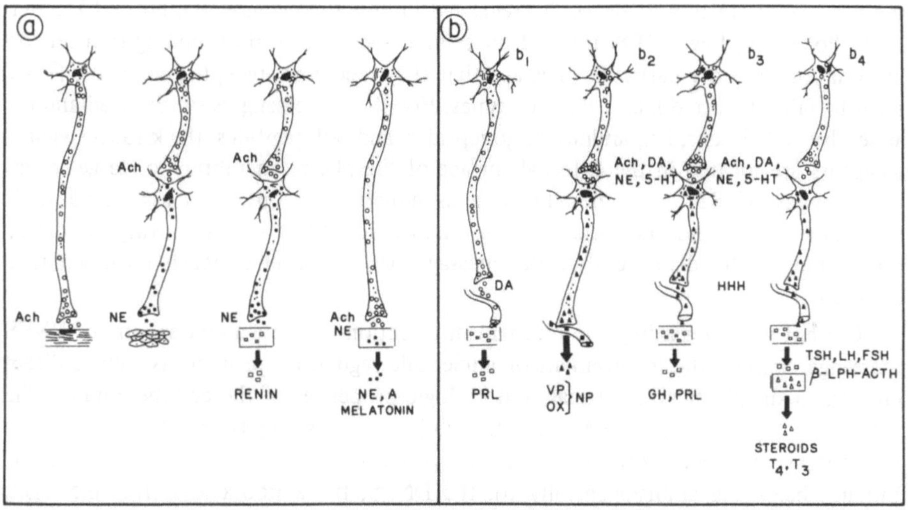

Fig. 1 a, b. This figure represents from the simplest to the most complex, the established morphofunctional signal processing organization of neurohumoral information transfer. **a** the orthodox neurocrine mechanisms; **b** the more recent neuroendocrine variants which take in account the hemal components in signal processing and which are more congruent with the concept of neuromodulation. For further details, see Table 2 and see text

Abbrevations see footnote p. 38

system which, through a hierarchically ordered family of extracellular chemical messengers, subserves a major homeostatic role in the organism. However, it is obvious that this process of communication encompasses not only an ample number of messages or sentences (see Zwick 1977): amino acids, amines, oligo- and polypeptides, as well as some lipid and modified amino acids; but also a large vocabulary: *yes-no,* enabling (get ready) – *go/stop-if;* and a miscellanea of morphofunctional arrangements (Fig. 1) to convey and encode-decode their chemical commands. Thus, based on some of their characteristics, i.e., physicochemical properties, the extracellular route they follow to reach their final destination, or the time course of their action, certain general operational distinctions for these informational molecules seem to emerge.

As depicted in Fig. 1 and summarized in Table 2, the various neurocrine paracrine messages are dispatched along distinct channels and they usually travel short distances to reach their effector cells. In contrast, the neuroendocrine groups of signals use a semiprivate neurohemal and/or neurohydrocrine route whereas the endocrine messages are released into the general systemic circulation. Thus, the neuroendocrine-endocrine signals are delivered indiscriminately to target and nontarget cells. In connection with these morphofunctional arrangements it must be emphasized that current experimental evidence indicates that a given message can be delivered by using one or more of these extracellular routes. On the other hand, cells that are targets for a particular message contain specialized molecules, named receptors, to specifically recognize that particular signal and decode the encoded sentence in such a way that the appropriate response ensues. As reviewed in detail elsewhere (Baxter and Funder 1979; Pollet and Levey 1980), based on cellular localization post-binding transfer for information, and the particular message that they recognize, receptor molecules (RM) seem to fall into three major distinct types. For those messengers synthesized and released by DNES cells, i.e., amines, oligopeptides, and polypeptides, the known specific receptors are located in the external surface of the plasma membrane of target cells. These types of RM are generally large asymmetric glycoproteins associated with membrane lipids and their constant of dissociation (Kd) seems to vary in direct proportion to the distance that the message has to travel to reach its target (see Table 2).

Current views regarding the mechanism of action of these ample spectra of DNES messenger implies the involvement of nucleotide regulatory proteins as well as direct changes in the physical, electrical, and biologic properties of the cell membrane. The other two types of known RM are intracellular proteins, highly specific to recognize the sentence encoded by those messengers synthesized and secreted by cells driven by, but not belonging embryogenically to, the DNES, the amino acid-derived messages, i.e., thyroid hormones and lipid-derived messages, i.e., steroid hormones. At least for the latter the RM is localized initially in the cytoplasmatic compartment of the cell and after binding and some conformational change, the message is translocated into the nucleus. There is ample evidence suggesting that these third order endocrine messengers regulate the expression of the genomic information and the operation of the biosynthetic machinery in practically all the cells of the organism (for review see: O'Malley et al. 1979; Baxter et al. 1979).

It is under this wide scope of neuroendocrine generated intercellular messengers that, instead of merely providing the reader with a review of endocrine and peptide studies, in this chapter we will discuss in a critical manner the possible role played by a group of these neurohumoral signals during the sleep-waking cycle. In doing so we will focus primarily on papers and/or reviews published in the last 3 years.

Table 2. Some features and operational distinctions between neurocrine, neuroendocrine, and endocrine communication

	Neurocrine	Neuroendocrine	Endocrine
Primary input			
Route	Nervous pathway	Hematic and nervous pathways	Hematic route
Locus of arrival	Synaptic terminal	Synaptic terminal Semiprivate distribution	Widespread distribution
Signal	"Neurotransmitter"	"Hormonal" and "neurotransmitter"	"Hormone," metabolic substrates, other blood-borne substrates
Time course	Ultrashort (milliseconds–seconds)	Short (seconds–minutes)	Long (minutes–hours)
Output			
Sentence (see Zwick, 1977)	Low molecular weight substances (100–300 daltons) Amino acids and amines Oligopeptides?	Low molecular weight substances (500–2,000 daltons) Amines and oligopeptides	Wide range of molecular species amines, polypeptides, lipid, and modified amino acids
Vocabulary	Rigid	Ample	Rich
Synthesis	Cytoplasmic nerve ending Microsomal	Microsomal nerve ending Ribosomal	Microsomal, cytoplasmic Ribosomal
Packing storage	Dense core vesicles	Post-translational modification Dense core vesicles Granules (<100 nm)	Post-translational modification Granules (100–600 nm)
Transport	Cytoskeletal involvement	Cytoskeletal involvement	Cytoskeletal involvement
Release distribution	Synaptic terminal. Ca^{2+} dependent Local contiguity	Neurohemal Circumventricular Ca^{2+} dependent Synaptoid arrangement Local-semiprivate	General circulation Ca^{2+} dependent systemic biochemical coding
Actions			
Readout	Ionic dependent Membrane interaction	Membrane receptor coupled to nucleotide regulatory proteins	Membrane, cytoplasmic, and nuclear receptors
Kd	Low (10^{-6}–10^{-8} M)	High (10^{-8}–10^{-12} M)	Very high
Onset	Immediate (seconds)	Rapid (minutes–hours)	Slow (minutes–hours–days)
Inactivation	Specific enzyme. Reuptake	Specific and non-specific enzymes	Specific and non-specific enzymes. Uptake
Extinction	Rapid with removal of messenger (minutes)	Slower decline (hours)	Slow (minutes–hours–days)

2 The Morphofunctional Framework

2.1 Hypothalamopituitary Pathways and Messages

Current experimental evidence (for review see: Pearse and Takor 1976; Polak 1979; Pérez-E. and Valverde-R. to be published) suggests that both the glandular (Rathke's pouch) and the neural (floor of the diencephalon) portions of the hypothalamo-pituitary unit derive from specialized thickenings – ventral neural ridges – of the neuroectoderm, thus forming part of the DNES.

The advent of histochemical and more refined biochemical and immunocytochemical techniques has revealed a widespread distribution of several putative neurohumoral messengers (Table 3) in DNES cells throughout the vertebrate nervous and gastroenteropancreatic systems. For the purpose of this chapter only those neuroendocrine cells from the hypothalamopituitary unit will be described. The whole area including the biochemical methodology, the regional and anatomic localization, and the ontogenic, metabolic, and neurophysiologic aspects, has been the subject of recent and extensive reviews elsewhere (Barnea et al. 1977; Vale et al. 1977; Terry and Martin 1978; Gainer 1977; Hökfelt et al. 1978a, 1978b; Elde and Hökfelt 1978, 1979; Moss 1977; Bloom 1980; McKelvy 1977, to be published; McKelvy and Epelbaum 1978; Renaud et al. 1978).

2.1.1 The Magnocellular Hypothalamopituitary System

This system is composed of the classic neurosecretory cells of the supraoptic (SON) and paraventricular (PVN) nuclei which are responsible for the synthesis and release of oxytocin (OX) and vasopressin (VP) and their carrier proteins, the neurophysins (Sachs et al. 1975; Ochs 1977; Zimmerman and Antunes 1976; Gainer 1977; Defendi and Zimmerman 1978; Brownstein et al. 1980). In addition, recent neurophysiologic studies have also indicated the presence of this type of neurosecretory cell in the periventricular region and the arcuate nucleus (Renaud et al. 1978). Similarly, the immunocytochemical studies have revealed VP cellular groups in the suprachiasmatic nucleus (Vandesande et al. 1975) and in the triangular nucleus of the septum (Weindl and Sofroniew 1976). Furthermore, some perikarya of the PVN exhibit a high concentration of estrogen receptors as well as positive immunoreactivity to gastrin-CCK (GAS-CC) enkephalins (ENK_S) and angiotensin (ANG-II) (Hökfelt et al. 1978b; Elde and Hökfelt 1979; Stumpf 1970; Rossier et al. 1980; Vanderleghen et al. 1979).

The primary afferent connections to the magnocellular nuclei arise, as confirmed by electrophysiologic studies (for review see: Renaud et al. 1978, 1979b), from the septum, amygdala, midbrain reticular formation, and hippocampus. Recently (Khachaturian and Sladek 1980) it has been shown that both nuclei receive a rich catecholaminergic (CA) input which may be ontogenetically dependent on a putative neurotrophic role exerted by the magnocellular neurons. By using anterograde tracing techniques, a functional connection between the SON and some circumventricular structures, particularly the subfornical organ (SFO), has been recently suggested (Miselis et al. 1979). It must be emphasized that the SFO is at the moment the more

Table 3. Some putative intercellular messengers of neuroendocrine origin

Compound	Molecular weight (daltons)[a]	Compound	Molecular weight (daltons)[a]
Glycine (Gly)	75	Bombesin[b]	1,620
γ-Aminobutyric acid (GABA)	103	Neurotensin[b]	1,673
Acetylcholine (ACH)	146	Melanotropin (MSH)	2,000
Noradrenaline (NA)	146	Secretin[b]	3,056
Dopamine (DA)	152	Vasoactive intestinal polypeptide (VIP)[b]	3,381
Serotonin (5-HT)	176	Endorphin (β-END)	3,425
Adrenaline (A)	183	Glucagon	3,485
Carnosine	226	Gastrin (GH)[b]	3,839
Thyroliberin (TRH)[b]	362	Cholecystokinin (CCK)[b]	3,918
Enkephalin (Leu and Met-ENK)	556,574	Corticotropin (ACTH)[b]	4,500
Oxytocin (OX)	1,007	Gastric inhibitory polypeptide (GIP)[b]	5,104
Angiotensin (ANG-II)	1,046	Insulin[b]	5,734
Vasopressin (VP)	1,084	Lipotropin (β-LPH)	9,948
Luliberin (LHRH)	1,181	Somatotropin (GH)	22,000
Somatostatin (SRIH)[b]	1,342	Prolactin (PRL)	22,000–23,000
Substance P[b]	1,348	Thyrotropin (TSH)	28,000

a According to the structural formula
b Peptides common to the nervous and the gastroenteropancreatic system

sensitive site in the brain for the induction of drinking behavior by ANG-II, and that the microiontophoretic administration of the peptide increase the firing rate of the SFO and the organum vasculosum of the lamina terminalis (OVLT). (Epstein 1978, 1980; Phillips et al. 1979; Fitzsimons to be published).

These studies have also confirmed that the amyelinic axons of both the SON and PVN neurosecretory cells project and descend toward the neural lobe of the pituitary gland. This projection constitutes the now classic peptidergic supraopticoparaventricular (SOPV) neurohypophyseal tract or pathway, through which the system operates as a *first order neuroendocrine system.* As depicted in Fig. 1 b_2, its morphofunctional arrangement allows the system to deliver its chemical signals to terminal target cells such as those in the kidney (collecting ducts) or the mammary gland (myoepithelial cells). There is also recent evidence demonstrating an ENK-ergic neurohypophyseal projection that may regulate the secretion of both nonapeptides (Rossier et al. 1980). Further, these immunocytochemical probes have revealed that the PV neurosecretory cells send projections to the external layer of the median eminence (Zimmerman 1976; Antunes et al. 1977; Defendi and Zimmerman 1978; Brownstein et al. 1980). The PV-tuber-infundibular pathway conveys the nonapeptide VP and its neurophysin (NP) into the semiprivate portal plexus (neurohemal route) to reach its putative adenohypophyseal effector cells, probably corticotrophs. Thus, in this instance the magnocellular hypothalamopituitary system seems to operate as a *third order neuroendocrine system* (Fig. 1 b_4) as suggested by the changes exhibited after the peripheral (third order) hormonal output has been manipulated (Zimmerman 1976; Defendi and Zimmerman 1978; Elde and Hökfelt 1978).

Besides these two pituitary projections, some fibers of the magnocellular system travel to extrahypothalamic areas. Certain PV neurosecretory cells seem to reach and "innervate" the choroid plexus at the level of the lateral ventricles and the OVLT (Defendi and Zimmerman 1978; Brownfield and Kozlowski 1977; Kozlowski et al. 1978). Recently (Nilaver et al. 1980), a dense VPNp and OXNp pathway arising from the PVN has been observed to project into the medulla and spinal cord. These anatomic observations agree with neurophysiologic and autoradiographic studies (Smith et al. 1980). The terminal field includes the locus ceruleus, the nucleus of the tractus solitarii, the dorsal motor nucleus of the vagus, the intercalatus and commissural nuclei, and the substantia gelatinosa. These unexpected findings suggest a greater functional role for the magnocellular hypothalamic system than previously considered. The hypothalamochoroid pathway may represent an alternate route to deliver neuron derived messengers, which as proposed by Roussy and Mosinger in 1935 may correspond to the process named "encephalohydrocrine." These VP-ergic fibers may regulate the formation and/or hydrolectrolyte balance of the cerebrospinal fluid (Schultz et al. 1977). Similarly, the medulla and spinal cord projections may be involved in modulating the afferent nociceptive or stressful input to the system ascending from the brain stem (Nilaver et al. 1980), as well as the cardiovascular responses observed during different autonomic neuroendocrine functional states (Cohen and Cabot 1979; Smith et al. 1980; Loewy and McKellar 1980). Although uncertain, the rest of the magnocellular extrahypothalamic projections strengthen the proposed role of VP and OX in certain behavioral patterns (De Wied and Versteed 1979).

2.1.2 The Parvicellular Hypothalamopituitary System

For purposes of the present review, this term refers to a heterogeneous and scattered population of hypothalamic neurosecretory cells of both aminergic or peptidergic nature which by immunocytochemical and/or electrophysiologic criteria send terminal endings to the palisade zone of the median eminence and/or to some other intra- or extraphypothalamic regions excluding the neural portion of the pituitary gland.

The major intrahypothalamic aminergic groups are the dorsal (groups A_{11} and A_{13}) and ventral (groups A_{12} and A_{14}) catecholaminergic systems. There is also a large group of neurons at the level of the mammillary bodies and extending to the mesencephalon and ventral tegmental area (groups A_{10}). Excluding dopamine (DA), so far there is no conclusive evidence for intrahypothalamic serotonin (5-HT), noradrenalin (NA), or adrenaline (A) cell bodies. Thus, the major aminergic afference to the hypothalamus seems to arise from the lower brain stem (Fuxe et al. 1978 a,b); Hökfelt et al. 1978a, 1978b). The perikarya of the so-called tuberoinfundibular dopaminergic system (TIDA) are located in the arcuate nucleus (ARC) and the periventricular area (Hökfelt et al. 1978b). Recently with the aid of the simultaneous histofluorescence-immunocytochemical techniques the putative neurocrine and neuroendocrine role of TIDA-terminals has been strengthened. The median eminence (ME) exhibits in the outermost 6–10 μm contact zone to the perivascular space of the portal vessels (layer 1) a dense reticulum of fine sized dopaminergic fibers, whereas the adjacent internal lamine (20–40 μm) contains the densest concentration of gonadotropin-releasing hormone (GNRH) varicosities (McNeill et al. 1980). As depicted in Fig. 1 b_1 and b_3, this close juxtaposition of TIDA terminals to portal capillaries and to distinct GNHR-positive fibers provides supportive morphological evidence for those studies showing high levels of DA in portal blood which inversely correlate with serum prolactin levels, as well as for its direct role as a prolactin release-inhibiting factor (PIF). In this particular situation the system seems to operate as a *modified first order neuroendocrine system.* Similarly the interrelationship between TIDA-fibers and the intermingled GNRH varicosities, particularly at the lateral portions, supports the suggested axo-axonic regulatory role played by the amine on the secretion of this and other hypophysiotropic peptides (Ben-Jonathan et al. 1977; Chihara et al. 1979; Krulich 1979; Fuxe et al. 1978a; Neill 1980; McLeod 1980; Gudelsky and Porter 1980).

Furthermore, a recent study combining autoradiographic and fluorescence histochemical techniques in the same tissue section has revealed a close anatomical interrelationship between CA-containing neurocytes and sex steroid hormones (Heritage et al. 1980). The localization pattern of these hormones was not sex-dependent but steroid-dependent. Both estradiol (E_2) and dihydrotestosterone (DHT) labeled noradrenergic containing perikarya were found in the brainstem (groups A_5 to A_7), as well as in dopaminergic cell bodies from ARC and the periventricular area (group A_{12}). In contrast, E_2 but not DHT, was found in labeled NA-ergic medullar neurocytes (groups A_1 and A_2), whereas DHT was concentrated in NA-ergic cells located in the dorsolateral corner of the fourth ventricle (group A_4). In addition, dorsomedial, para- and periventricular nuclei which contain target neurocytes to both steroids, also receive a rich network of CA-ergic endings. A recent immunocytochemical study

(Reichlin et al. 1980) demonstrates a close antigenic relationship between "estrophilin" derived from peripheral estrogen target cells and that contained in diencephalic neurons.

The peptidergic component of the parvicellular hypothalamopituitary system encompasses a growing family of intercellular messengers which seem to convey information throughout neurocrine (paracrine), neuroendocrine, and endocrine mechanisms (see Fig. 1 and Table 2). There is a large group of oligopeptides which are common to the central and peripheral division of the DNES, whereas the polypeptides seem to be restricted to the former and the placenta (for review see: Pearse 1977; Pearse and Takor 1976; Grossman 1979; Pérez-E. and Valverde-R. to be published).

Perikarya containing the tripeptide thyroliberin (TRH) have been localized in the dorsomedial hypothalamus, in the parvicellular portion of the OVN, and at the level of the anterior periventricular area (Hökfelt et al. 1978b; Elde and Hökfelt 1979; Terry and Martin 1978). The major TRH-ergic tuberoinfundibular projection seems to be provided by neurons from the periventricular zone. These fibers extend to the external layer of the ME and are highly concentrated on its central portion where they appear to about on the capillary loops of the portal plexus (Hökfelt et al. 1978b; Elde and Hökfelt 1978). There is a rich intranuclear TRH-ergic interconnection, particularly at the level of the dorsomedial hypothalamus and PVN, as well as an ample internuclear arrangement encompassing the ventral preoptic area and the suprachiasmatic (SCN) and ventromedial (VMN) nuclei. There is also a prominent TRH-fiber component in more distant extrahypothalamic areas, particularly the brain stem and the ventral horn and intermediolateral coll column of the spinal cord (Elde and Hökfelt 1978; Hökfelt et al. 1978b). This intricate network of nontuberoinfundibular TRH-ergic pathways may represent, at least in part, the anatomic substratum for the well known neurotropic effects exerted by the tripeptide (Guillemin 1978; Moss 1979; Renaud et al. 1979 a and b).

Compared with the rest of known hypophysiotropic oligopeptides GNRH seems to have the most limited distribution. Although still controverted (Flerko 1980) there is practically general agreement that the intrahypothalamic neurons containing the decapeptide are located in the ARC and SCN whereas the extrahypothalamic GNRH-positive perikarya have been found at the level of the anterior preoptic area and the diagonal band of broca near the OVLT (Stemberger and Hoffman 1978; Knigge et al. 1978; Ibata et al. 1979). The ultrastructural features of these GNRH-containing neurons have been recently reported and they seem to be morphologically similar despite their different location and projections (Kozlowski et al. 1980). In addition, this report has revealed that GNRH perikarya do not seem to receive axo-somatic synaptic terminals. The major known projections of these GNRH-containing neurocytes include two circumventricular-neurohemal regions: the external layer of the ME including its lateral lips at the level of the tuberoinfundibular sulcus and the OVLT (Hökfelt et al. 1978a; McNeill et al. 1980). Therefore these anatomical observations suggest that some parvicellular oligopeptidergic neurons may use a semiprivate route, neuroendocrine and encephalohydrocrine, to convey their message. The major extrahypothalamic GNRH projection areas include the septum and amygdaloid complex as well as the mesencephalon. These terminals may be involved in the putative nonhypophysiotropic role that GNRH seems to play in the expression of sexual behavioral patterns (for review see: Moss et al. 1979).

Within the hypothalamus the major group of neurocytes containing immunoreactive somatostatin (SRIH) has been observed at the level of the anterior periventricular area. There is also a cellular group in the entopeduncular nucleus-zona incerta region and in several extrahypothalamic areas including the amygdaloid complex, hippocampus, and cerebral cortex. There is clear evidence of a rich tuberoinfundibular SRIH-ergic pathway arising from the periventricular cell group whose nerve endings form a dense network all over the ME. Although several intra- and extrahypothalamic areas, such as the ARC, VMN, SCN, and the OVLT, or the caudate nucleus, olfactory tubercle, and some lower portions of the brain stem, exhibit positive SRIH nerve terminals, the origin of these endings is unknown (Hökfelt et al. 1978b; Elde and Hökfelt 1978, 1979).

Among the newcomers to the heterogeneous group of intrahypothalamic parvicellular neurosecretory systems we find the so-called pro-opiomelanocortin peptides. We will only briefly review some pertinent aspects. Recent work (Pelletier and Desy 1979; Pelletier and Leclerc 1979; Watson et al. 1980) has shown the existence of β-lipotropin (β-LPH) and β-endorphin (β-END), corticotropin (ACTH) positive neurocytes, within the hypothalamic area, whereas melanocyte-stimulating hormone (α-MSH) neurocytes are confined to the ARC and the zona incerta (Guy et al. 1980). Although the distribution pattern for these peptides is similar in general terms, the use of specific antisera and consecutive sections has shown that they are contained in different cell bodies. Neurocytes containing immunoreactive ENKs have been observed within the ARC, VMN, premammillary, and PVN and SON nuclei (Hökfelt et al. 1978b; Fuxe et al. 1978; Elde and Hökfelt 1979; Rossier et al. 1980). In addition to the tuberoinfundibular projection of this system which may operate throughout neuroendocrine mechanisms, the nerve fibers of the proopiomelanocortin arcuate-neurosecretory cells seem to constitute a diffuse neurocrine interconnecting intra- and extrahypothalamic system. The intrahypothalamic arcuate α-MSH system and its projective internuclear field: medial preoptic, anterior, periventricular, dorsomedial, PVN, and ME, exhibit a significant circadian rhythm (O'Donohue et al. 1979). Recently Watson et al. (1980) have shown the existence of a reciprocal morphological interrelationship between the NA-locus ceruleus system and the β-END/α-MSH-intrahypothalamic system which may represent the anatomic substratum for some of the ample spectrum of biologic effects exhibited by this group of peptides (for review see: Beaumont and Hughes 1979; Van Vugt and Meites 1980; Cicero 1980; Dupont et al. 1980).

A similar functional interrelationship seems to exist at the level of the adrenal medulla where the presence of ENKs in the chromaffin granules has recently been demonstrated (Viveros et al. to be published; Lewis et al. 1980). Furthermore, these peripheral endogenous opioid peptides are released from the gland into the general circulation, thus suggesting that they may play an unknown endocrine function (Hexum et al. 1980). In addition to these immunocytochemical observations there is also recent evidence suggesting that the intrahypothalamic pro-opiomelanocortin group of peptides may be synthesized *in situ* (Liotta et al. 1979; Scherrer et al. 1979).

There is also conclusive evidence indicating the presence of "purely polactin-positive" neurocytes at the level of the ARC, VMN, and premamillary nuclei as well as in the SON and PVN which seem to be independent of the presence of the pituitary gland and/or the corresponding pituitary function (pregnancy and/or lactation). The immunoreactive intrahypothalamic nerve terminal plexa containing prolactin (PRL) was observed within the periventricular area and arcuate nucleus. The main extra-hypothalamic projections include the anterior and medial preoptic areas, the medio-basal olfactory cortex, the medial amygdaloid nucleus, the locus coeruleus, and the motor nucleus of the vagus and hypoglossal nerves. (Fuxe et al. 1978; Hökfelt et al. 1978b; Fuxe et al. 1979; Toubeau et al. 1979 a, b). These morphological observations add further support to those studies (for review see: Weiner and Ganong 1978; Annunziato 1979; Krulich 1979; Gudelsky and Porter 1980), indicating that there is a feedback relationship between circulating PRL levels and the activity of the TIDA system, which seems to involve the synthesis of the rate-limiting enzyme tyrosine hydroxylase (Johnston et al. 1980). It must be emphasized that this type of direct feedback on the hypothalamus (also called internal or short-loop feedback) seems to operate also for the rest of hypothalamopituitary messengers secreted via the semi-private neurohemal route provided by the portal system. There is conclusive evidence supporting the existence of an upward or retrograde blood flow in the pituitary stalk reaching a subependymal plexus immediately below the ependymal surface of the third ventricle (for review see: Flerko 1980; Bergland and Page 1980). However, some findings argue against this hematic route as the only mechanism explaining the presence of pituitary peptides within the hypothalamus. Both PRL- and ACTH-like hypothalamic peptides are not affected by hypophysectomy, pregancy, lactation, adrenalectomy, and water deprivation, and both seem to exhibit a postnatal ontogenetic pattern (Toubeau et al. 1979 a, b). The pro-opiomelanocortin family of neuropeptides is synthesized by the hypothalamic tissue (Liotta et al. 1979; Scherrer et al. 1979).

Several other "classic" pituitary hormones: thyrotropin (Guansing et al. 1977), somatotropin (Pacold et al. 1977), and luteotropin (Emanuele et al. 1979), as well as several gastroenteropancreatic peptides, i.e., substance P, vasoactive intestinal peptide, cholecystokinin, and insulin have been detected within the hypothalamus and/or other cerebral areas. Although the elucidation of the putative role of most of these peptidergic messengers is still largely in its initial stage, they may influence the function of the hypothalamopituitary unit through neurocrine and/or neuroendocrine mechanisms (Elde and Hökfelt 1979; Fuxe et al. 1979; Vijayan et al. 1979; Van Houten et al. 1980; Zieglgansberger 1980).

From the above it is evident that the complexities of neurohumoral secretions, their interactions, and their overlapping anatomic localizations make their interpretation in terms of any physiologic function an inordinately difficult, if not impossible, task, given the present day scattered knowledge. For this reason, the reader should conceive these hormones perhaps as functional oscillators which give off special signals at given periods in time. Sleep lends itself to such mechanisms.

3 The Chronobiologic Framework

3.1 Sleep and Known Neuroendocrine Factors

As recently stated (Aschoff 1979) the rhythmic structure of biologic systems can be considered a principle complementary to that of homeostasis. Current experimental evidence (for review see: Aschoff 1979; Rusak and Zucker 1979) supports the concept that biologic systems incorporate into their own organization the temporal structure of their environment. These approximate copies of the external-geophysical chronological program provide the organism with the so-called biological clocks which allow the system to be prepared to cope for programed or predicted environmental changes.

Sleep and wakefulness (SW) stand as a paradigm of these major rhythmic adaptive biologic mechanisms, representing a cyclic physiologic diad originating primarily from structures and processes within the central nervous system (CNS). Furthermore, the operation and maintenance of this endogenous self-sustained rhythmicity which persists even in the absence of environmental cues (free-running conditions) seems to represent the accurate synchronism of an ontogenetic program or neural clock to the periodicity of a major geophysical environmental cycle: the terrestrial solar day. Thus, the SW cycle is an active internally programed circadian rhythm, i.e., it is a programed endogenous oscillation in close (circa) correspondence – entrained – with a period of a day (diem). In addition, a major part of the S-W neural program seems to involve a complex cascade of coupled controlling or regulatory neurocrine and/or neuroendocrine-endocrine signals generated and/or integrated by the hypothalamopituitary unit and other CNS components of the DNES. There is a growing body of experimental evidence (for review see: Moore 1978, 1979; Rusak and Zucker 1979) indicating the existence at the level of the anterior hypothalamus of a putative neural photoreceptor complex which may serve as a major pacemaker for the generation of several circadian rhythms including the S-W cycle. The SCN is envisaged at the moment as a pivotal circadian oscillating unit probably composed of several coupled oscillators (Moore 1979; Abe et al. 1979), one of which receives a direct visual (light-dark) input through the retinohypothalamic (RTH) pathway thus entraining the circadian endogenous program(s) of the coupled oscillator(s). Besides this RTH afference that conveys photic information, the SCN receives several other major inputs of aminergic and peptidergic nature. There is an ascending serotoninergic projection from the midbrain raphe nuclei (nucleus centralis superior) and an indirect visual projection arising from the ventrolateral geniculate nuclei. Similarly, the SCN is surrounded by a dense capsule of catecholaminergic axon terminals (Moore 1979; Rusak and Zucker 1979), and, as previously mentioned (see Sect. 2), there is evidence indicating a TRH and SRIH-ergic SCN terminal field derived from the corresponding intrahypothalamic parvicellular neurocytes. On the other hand, as already mentioned, the SCN contains at least two types of neurosecretory cells, a parvicellular group of VP-positive perikarya and a group of GNRH immunoreactive neurons. The efferent connections of the SCN are poorly defined as yet, but they seem to include at least two major terminal fields. There is an intraphypothalamic direct periventricular-tuberal projection that appears to terminate in the arcuate nucleus and the ME. There is also an extrahypothalamic indirect brainstem spinal cord projection arising from the lateral hypothalamic area which receives a projection from the ventral tuberal area (for review see: Moore 1979).

Furthermore, by preparing an "hypothalamic island," it has been shown that circadian multiple unit activity was preserved within the island containing the SCN but was lost in the rest of brain sites tested, i.e., dorsal and medial raphe nuclei, caudate nucleus, locus ceruleus, mesencephalic reticular formation, and other (Inouye and Kawamura 1979). These findings parallel those derived from lesioning the SCN in both adult and neonatal animals and showing arrythmic profiles in diverse circadian variables (for review see: Moore 1979; Abe et al. 1979; Moore 1979). Similarly they agree with the finding of a circadian rhythm on the uptake of ^{14}C-deoxyglucose which seems to be unique for the SCN (Schwarz and Gainer 1977). Finally, there is a growing body of experimental evidence suggesting that the SCN rhythmic activity can be affected by endocrine signals derived from the pituitary gland, the pineal gland and the gonads (Rusak and Zucker 1979; Zucker et al. 1980). This type of "phasing influence" may well be of major physiologic importance for maintaining appropriate synchronization of the internal mileu as well as coordination of the organism with the external environmental demands. Thus, although firm conclusions cannot yet be drawn, these data strongly support the putative role of the SCN as a primordial endogenous set of mutually coupled circadian oscillators with important hormonal regulatory influences in mammalian species.

3.1.1 Sleep and Hormones

As reviewed in detail elsewhere (Boyar 1978a, b; Weitzman et al. 1979; Weitzman 1980), during the past 10 years a large body of experimental information has accumulated concerning the phenomenological description of the temporal association exhibited by several neuroendocrine signals and the S-W rhythm. However, these phase-relationships should not be interpreted as evidence in favor of a simple cause-and-effect dependence of a given neuroendocrine variable on sleep or vice versa. On the contrary, current available information suggests that, many, if not all these rhythmic processes probably represent separate circadian and/or ultradian oscillators passively driven and/or synchronized by the S-W rhythm, and that this functional linkage occurs with variable intensities. In the following paragraphs we will briefly summarize the temporal pattern of several of these neuroendocrine rhythms in human subjects under normal 24-h S-W conditions, anticipating that these studies give no clues as to the role of hormones in the mechanisms of sleep.

The neurosecretory activity of the magnocellular hypothalamopituitary system stands as the classic-orthodox "glandular" function of the brain. Furthermore, as previously mentioned (see Sect. 2) there is a large body of information regarding the synthesis, transport, secretion, distribution, biologic effects, and mechanisms of action of these groups of neuron-derived messengers. However, the information concerning their circadian variations is scarce. As far as we know there are only two reports describing plasma VP episodic secretion and nocturnal increase without significant relation to sleep stages, plasma osmolality, or hematocrit (Georg et al. 1975; Rubin et al. 1975).

On the other hand several studies (for review see: Seif and Robinson 1979) in which the environmental temperature was modified indicate the possibility that VP secretion in some animal species may exhibit infradian – seasonal – variations in response to long–term temperature fluctuation, whereas the short–term fluctuations may influence the daily secretory pattern. Similar studies in man are lacking. In addition, it is well known that the intravascular distribution of blood secondary to postural changes elicits concomitant changes in plasma VP levels. Thus the diurnal variation of this peptide may be associated to the physiologic hemodynamic changes occurring during sleep. Furthermore it seems reasonable to suggest that part of the nocturnal cardiovascular changes may be secondary to neurocrine information probably conveyed through the recently described descending PVN pathway which projects to the midbrain and spinal cord (see Sect. 2). Although there is no information regarding the temporal daily pattern of OX secretion, a clear relationship between plasma estrogens, LH, and the so-called estrogen-stimulated neurophysin or OX-Np has been documented in primates and women during the mid-menstrual cycle (Amico et al. 1980). As previously mentioned (see Sect. 2) estrogen receptors have been found in the PVN, and the steroid is able to increase the spontaneous firing rate of these magnocellular hypothalamic nuclei (Negoro et al. 1973).

A large body of information indicates that the daily secretory pattern of GH and PRL is a direct sleep-dependent ontogenic programed rhythm (for review see: Boyar 1978 a, b; Takahashi 1979; Parker et al. 1979; Frantz 1979). That is to say, their periodicity is tightly, and over an ample range of physiologic and experimental situations consistently, linked or entrained to the frequency of the S-W cycle. Thus, whereas GH sleep-related secretion usually occurs during the first sleep cycle and is temporarily associated with the appearance of the initial SWS stage or high voltage delta activity (Takahashi 1979); PRL on the other hand exhibits a characteristic steady progressive increase, with recurrent peaks during non-REM sleep and nadirs during REM stages (Parker et al. 1974; Frantz 1979). However, as reviewed elsewhere (Parker et al. 1979; Takahashi 1979; Wagner and Weitzman 1980; Frantz 1979; Solis et al. 1979; Valverde-R. et al. 1980), there is a large group of pharmacologic and pathologic conditions in which GH and to a lesser entent PRL secretion become dissociated from the concomitant sleep events previously described.

Taken together these results support the notion of separate neuroendocrine oscillators which are tightly bound and passively driven by the circadian S-W rhythm. However, the physiologic meaning of these sleep related events will remain uncertain until we learn more about the biologic role played by GH and PRL. These two versatile single chained polypeptides synthesized by DNES cells within the hypothalamic-pituitary unit are considered to have evolved from a common ancestral molecule (for review see: Bentley 1976). Both represent a paradigm of the so-called second neuroendocrine signals exerting their ample, and in certain situations, overlapping spectra of biologic effects on a similarly ample population of cells. Although GH is the primary somatic growth factor in humans and lower animals, there is an expanding family of proteins that regulate cell proliferation and function of specific tissues. However, and in contrast to these group of polypeptides, GH does not seem to have a specific target tissue nor to regulate any particular function. On the contrary, GH exerts its wide range of metabolic and growth-promoting effects on a similar ample number of target cells by regulating the synthesis of a group of relatively low molecular weight polypeptides, the somatomedins, as well as by exerting a direct regulatory effect on DNA, RNA, and protein synthesis in some tissues (for review see: Roger 1979; Golde 1980). Similarly, PRL has been shown to exert around 85 different actions grouped in six major categories: osmoregulation, growth development, metabolic, reproduction, ectodermal, and synergic-antagonic steroid and thyroid hormone effects (for review see: Nicoll 1980). The mechanism(s) by which any one of the many actions of PRL is carried out is not completely understood. They may vary according to the target cell involving in some instances specific rm on the plasma membrane that may be influenced by the ionic environment, and probably also some direct nuclear locus of action (for review see: Rillemal 1980).

There is conclusive evidence showing that the temporal organization of the hypothalamic-pituitary-adrenal cortex subsystem results from the oscillatory activity of an endogenous CNS pacemaker. There is a cluster of stimulatory-secretory events occurring just prior to and during the first hours of waking, without any particular temporal association with sleep stages. Further, several studies support the notion that, although somewhat independent, there is a reciprocal functional coupling between this proopiomelanotropin-cortisol pacemaker and the S-W driving oscillator (for review see: Weitzman et al. 1979; Mullen et al. 1979). There is also evidence suggesting that a similar operational arrangement may be involved in the temporal pattern exhibited by the hypothalamic-pituitary-thyroid subsystem. However, in contrast to the rest of the neuroendocrine signals, thyroid-stimulating hormone (TSH) exhibits its major secretion during the presleep period and/or just after sleep onset (for review see: Wagner and Weitzman 1980).

In the case of gonadotropins, particularly LH, a clear sleep-related secretion during the peripuberal stages in both sexes has been shown. These patterns seem to represent part of a programed CNS mechanism independent of the gonad (Boyar 1978 a, b).In addition the ultradian rhythmicity exhibited by LH in mature women is probably due to pulsatile discharges of LHRH from brain peptidergic neurons (Gallo 1980).

Evidently all these studies merely represent phenomenological descriptions which add little to the understanding of the role played by these hormones in the mechanisms of sleep. Furthermore, despite the fact that these neuroendocrine messages affect one or more of the following functions: (1) positive or negative feedback actions on the hypothalamopituitary secretory output, (2) positive or negative actions on neuroaminergic metabolic pathways, (3) inductive and/or organizing effects on the brain, (4) plastic and consolidative effects related to adaptive behavior or learning, (5) control of specific patterns of instinctive behavior, (6) general metabolic effects; there are practically no studies on the effects of hormones on sleep.

3.1.2 Hormones and Sleep

To the best of our knowledge the only sleep studies assessing the effect of hormones or some other peptides have been done using GH, TSH, SRIH, and arginine vasotocin (AVT). Moreover the experimental model offered by pathology has been practically neglected. As reviewed elsewhere (Valverde-R. et al. 1980; Valverde-R 1980), it has been found that at least for the case of both primary and secondary hypothyroidism, thyroid hormones may represent an important synchronizing and/or permissive signal to the occurrence of a normal S-W rhythmicity. Similarly, recent studies suggest that TRH may affect sleep rhythmicity throughout an independent nonendocrine mechanism (Jurado et al. 1979; Fernández-Guardiola personal communication). Furthermore, in accordance with this putative role of TRH, it has been recently suggested that both TRH and SRIH may vary in opposing directions at sleep-onset (Peters et al. 1980).

The most encouraging results in relation to an effect on sleep come from the observations that GH produces a dose-dependent increase of REM sleep in rats; cats; and humans (Drucker-Colin et al. 1975 a; Stern et al. 1975; Mendelson et al. 1980), while SRIH decreases sleep in rats (Havliceck et al. 1976). These studies are discussed more extensively later. TSH, on the other hand, has no effect on sleep (Drucker-Colin et al. 1975 a; Stern et al. 1975).

Although highly disputed, there is a possible direct sleep effect exerted by some neuropeptides. One such case is arginine vasotocin (a combination of the ring of OX and the side chain of VP), which is considered to be the "parent" or original ancestral molecule of the nine naturally occurring peptides of the family. Arginine vasotocin (AVT) has been found in all vertebrates, and probably also during fetal life in the mammalian group (for review see: Bentley 1976).

Using a differential bioassay approach it has been reported that AVT is found in the mammalian pineal gland (Pavel 1965; Pavel et al. 1977 a). These initial findings were confirmed using radioimmunoassay (RIA) (Rosenbloom and Fisher 1974) and immunohistochemical staining procedures (Bowie and Herbert 1976). However, more recent reports have failed to support these previous observations (Negro-Vilar et al. 1979; Dogterom et al. 1979). Furthermore, by combining high performance liquid chromatography and RIA procedures, a much lower value of bovine pineal AVT (3 pg/gland) than previously found has recently been reported. In contrast, sizeable amounts of VP and OX (300 pg/gland and 110 pg/gland) were detected (Fisher et al. 1980). Since AVT has been suggested to have a number of actions in the central nervous system unrelated to its antidiuretic role (Pavel et al. 1973; De Wied and Versteeg 1979) it will be important to resolve this controversy, since its unequivocal presence in the pineal gland of mammals in an obvious requirement for accepting its role in physiologic events.

One of the roles in which AVT has been implicated in sleep. For example, Pavel et al. (1977 b) have reported that extremely small doses of AVT (10^{-6} pg) corresponding to about 600 molecules, administered into the third ventricle of cats, induces prolonged periods of slow wave sleep. An additional interesting fact is that AVT completely abolishes REM sleep. This increase in slow wave sleep is not observed when AVT is incubated with trypsin, and is not mimicked by VP or OX. Unfortunately this study should have provided more sleep data, since in the only table reported the total amount of slow wave sleep does not change with treatment. From what can be determined in the text AVT seems to shorten the latency of slow wave sleep. In a subsequent study Pavel et al. (1979) reported that the effects of AVT on sleep are mediated by serotonin. This was substantiated by observations that AVT, but not VP or OX increases brain level of 5-HT and that administration of AVT + Fluoxetine, a specific and selective 5-HT uptake inhibitor (Fuller and Wong 1977), doubles the increase in SWS, while administration of AVT + methergoline, a 5-HT receptor blocking agent (Fuxe et al. 1978 b), eliminates SWS.

There are two recent studies, however, which have been unable to confirm the sleep-inducing properties of AVT (Mendelson et al. 1980; Tobler and Boberly 1980). The Mendelson study did nonetheless confirm the REM sleep-inhibiting properties of AVT. The main difference between these latter two studies and Pavel's is the fact that they used rats instead of cats. It is conceivable, then, that AVT may have differing effects depending on the species. Should this be the case for sleep, it would make the interpretation of AVT as a sleep-inducing peptide difficult. In conclusion it appears that the verdict on AVT as a sleep hormone should be withheld until further evidence becomes available.

3.2 The Search for Sleep Factors: The Sleep Peptides

The investigation on sleep peptides, which has centered around the search for sleep inducing substances, has a long history, but has been very slow in developing, mostly because it has had very few proponents. This is probably due to the fact that the isolation of a factor having a specific function represents a monumental task, given the enormous number of factors which are likely to exist in the brain. In addition there are other problems, mainly related to the adequacy of the experimental approach. It is not obvious a priori whether sleep-related or sleep-inducing substances will appear during wakefulness, at the onset of sleep, during sleep, or immediately after a REM period as recently suggested by Jouvet (1978). All these alternatives can impose important methodological constraints.

No good evidence exists for a substance elaborated during sleep whose levels differ between sleep and wakefulness. Even during prolonged wakefulness, measurement of presumed hypnogenic neurotransmitters such as norepinephrine (NE) and 5-HT has failed to detect changes in levels (Stern et al. 1971), perhaps because of increased utilization (Hery et al. 1970). This suggests that prolonged wakefulness does not induce the accumulation of "hypnotoxins," be they peptides or neurotransmitters. Even were there such evidence, there remains the dilemma of determining whether material obtained during sleep is the cause or the result of sleep. Hormones present similar problems.

Regardless of whether the search for sleep-inducing substances is done during wakefulness or during sleep, there remains the problem of where to search for them. Should the search be done in the ventricles, in plasma, or directly in brain tissue? If the latter is chosen, there remains the problem of where in the brain to search for them. If it is decided to search in the brain stem, which seems most reasonable given the available evidence (Moruzzi 1972), there remains the problem of deciding whether to look into the "waking system" or in the "sleep system," which brings us back to the question of when in the sleep-wake cycle to look.

In addition to these methodological questions, there are other important reasons that very few laboratories have attempted to identify sleep-inducing substances. First, there is an absence of a good and simple bioassay for determining the sleep-inducing activity of the biochemical fractions at the various stages of purification. Second, the length of time which must be invested in this approach, and the relatively poor prognosis of success. The final reason is that unfortunately, as we shall see below, the results so far of all the studies that have attempted to identify sleep-inducing peptides may merely be artifacts of the procedures.

It is clear that most studies related to the search for sleep-inducing substances have been, in one way or another, influenced by the pioneering experiments of Pierón. This investigator reported (Legendre and Pierón 1910, 1911, 1912) that serum or cerebrospinal fluid (CSF) obtained from sleep-deprived dogs induced a state of somnolence or "sleep" when introduced into the fourth ventricle of normal dogs. If these fluids were ultrafiltrated, dialyzed, or heated at 65 °C, their sleep-inducing properties were lost. Though not proposed at the time, these latter observations may have been anticipating the possibility that the "sleep-inducing substance" was of a peptidergic nature. At any rate Pierón's observations prompted him (1913) to suggest that during wakefulness the CSF accumulated a "hypnotoxin" with properties conducive to sleep and which dissipated during sleep.

Many years later, Schnedorf and Ivy (1939) repeated Pierón's experiments, observing behavioral sleep in 9 out of 24 dogs injected intracisternally with 8 ml CSF from dogs deprived of sleep for periods of 7–16 days. However, "sleep" was also observed in 4 out of 24 dogs receiving CSF from non-sleep-deprived dogs. Similar results were reported by Kroll (1933) with brain extracts of cats and rabbits previously put to sleep with drugs. This brain extract induced "sleep" in a recipient animal. The results of all these studies were complicated by the large volumes of CSF withdrawn and injected into the ventricles which caused increases in temperature and ventricular pressure. Also, "sleep" was judged by simple behavioral observation, since at the time, no objective quantitative measures were available. Without such measures it is difficult to distinguish sleep from responses to the stressful procedures used.

In the more modern versions of these experiments some of these factors have been corrected, i.e., volume factors. However, as we will see below, the assessment of sleep leaves much to be desired, which is unfortunate given that such assessment is relatively easy to do and universally utilized by sleep researchers.

3.2.1 Factor S

In what could be considered the most direct extension of Pierón's original observation, Pappenheimer and his group have carried out a series of studies under the theoretical framework that sleep is the result of substances that accumulate during wakefulness and can be found in the ventricles. In their first study (Pappenheimer et al. 1967), goats, chronically implanted with a cisternal tube, were deprived of sleep by an automatic device that provided electric shock and acoustic alarm to relaxing animals. At the end of the deprivation period the cisterna magna was punctured and CSF withdrawn at a rate of 0.1 ml/min for 5 h.

Cerebrospinal fluid from sleep-deprived and normal goats was injected into the lateral ventricles of chronically implanted rats at a rate of 3.3 μl/min for 30 min (0.1 ml total volume). Injections occurred in the late afternoon prior to the dark period, when rats show their highest level of activity. "Sleep" was measured behaviorally on the basis of locomotor activity monitored constantly by photocells connected to automatic counters. CSF from sleep deprived goats decreased locomotor activity of rats for several hours, while normal CSF had no effect. Since some pilot experiments in cats showed similar effects, the authors wrote: "the fact that fluid from sleep-deprived goats is active in cats and rats suggests that we are dealing with a humoral factor of general and fundamental importance to the sleep mechanism."

Unfortunately this study leaves much to be desired. First, the rats' ventricles were punctured at the time of each infusion, which was probably not only stressful but also involved a great deal of handling. It has long been known that handling is aversive to rats (Candland et al. 1962). The procedure may have resulted in periods of inactivity immediately following the treatment, thus decreasing mean locomotor activity. Injection of saline (i.p.) alone can reduce activity counts (Schnitzer and Ross 1960). Moreover, the authors only reported standard errors and ranges of activity counts for the preinfusion control, and not for the periods following treatment. Since the data are presented for 6- or 24-h periods, it is conceivable that the low mean is the result of very low activities in the periods immediately following injections. It would have been more convincing if the authors had compared hourly trends. On the other hand, since mean activity counts in rats given CSF from sleep-deprived goats were lower than those given normal CSF, the results hint at the possibility that the former CSF indeed contained a sleep-inducing humoral factor not present in the latter. However, the only way to ascertain whether this factor induces sleep is to determine it electrophysiologically. Unfortunately no such electrophysiologic recordings were made.

In subsequent studies this problem was partially corrected. Locomotor activity and cortical EEG were recorded in a few chronically implanted rats, and a good correlation between low activity and increases in EEG synchronization was reported (Fencl et al. 1971). Also, 0.1 ml CSF from a goat deprived of sleep for 48 h induced the greatest decrease in locomotor activity. It was therefore concluded that the concentration of the sleep-inducing factor in CSF peaks after 48 h of sleep deprivation.

Molecular sieving experiments using calibrated membranes indicated that the sleep-inducing factor had a molecular weight of less than 500 and it was named Factor S (Pappenheimer et al. 1975). During chromatography on Sephadex G-10, Factor S eluted just before sucrose. Reaction with fluorescamine indicated that 0.5 nmol of a peptide-like substance in this region was biologically active, although no correlation was reported between biologic activity and fluorescamine reactivity. Incubation of Factor S with Pronase destroyed its biologic activity leading to the suggestion that it was a peptide. The sleep-inducing properties of Factor S were not mimicked by 5-HT, γ-hydroxybutyrate, butyrolactone, GABA, glutamate, or cyclic 3′5′-AMP. The presence of Factor S at periventricular sites was suggested.

The ultrafiltrates were tested on locomotor activity in rats and on the EEG of rabbits bearing screw electrodes over the frontal and occipital cortices. 0.1 ml of purified concentrate of CSF from normal and sleep-deprived goats significantly decreased for 12 h nocturnal locomotor activity in rats in comparison to those receiving saline. Since a 20-fold concentration of CSF from non-sleep-deprived animals also produced an effect, it was suggested that Factor S is present in small concentrations even in normal animals. This observation indicated that one could forego the painful process of sleep deprivation. In rabbits, 0.1 ml of purified Factor S increased the relative amount of SWS by approximately 15%–20% for 4–5 h. Also, the mean integrated voltage of the EEG was increased relative to normal and was similar to that observed following sleep deprivation.

An important observation in this study was that brain extracts, particularly those from the brain stem, also appeared to contain Factor S. In the most recent report of this group (Krueger et al. 1978) Factor S was extracted from the brain and treated by partition chromatography and high voltage electrophoresis to yield an active product purified one million-fold. The concentration of Factor S in brain tissue appears to be of the order of 0.3 nmol/gram tissue, the effective sleep-inducing dose being about 0.15 nmol.

During the course of these studies, it appeared that CSF also contained a factor capable of inducing a long-lasting excitatory behavior. In contrast to Factor S, this excitatory Factor E appeared to exert its effects even if infused into the subarachnoid spaces, this suggesting receptors in cortical tissue. The presence of the so-called waking factors have been previously reported in cross circulation experiments in rabbits (Monnier and Hosli 1964) and in *encephale isolè* cats (Purpura 1956).

Factor S therefore appears to be a peptide, present throughout the brain, whose concentration increases as a result of sleep deprivation and which has sleep-inducing properties in several species. These observations could be taken to signify that sleep is the result of a delicate balance between the accumulation and utilization of a universal peptide. However, before accepting this postulation a few additional facts should be considered. First, Ringle and Herndon (1969) did not observe any effects on locomotor activity, behavior, or EEG patterns with CSF obtained from sleep-deprived rabbits. This might mean that sleep-deprived rabbits do not produce a sleep-inducing factor and that such a factor is not so universal. In all fairness, however, this study is not strictly comparable to Pappenheimer's since the procedure of extraction of CSF involved prior electrocution of the rabbits.

The most serious problem in Factor S studies is the absence of sleep-wake recordings. Granted that the authors have a relatively rapid bioassay, this phase should now be totally replaced by complete sleep electrophysiology including REM sleep and its concomitant phasic activities. Until such studies become available, one should be very sceptical of the significance of Factor S, and the consideration of Factor S as a sleep inducing peptide withheld.

3.2.2 Delta Sleep-Inducing Peptide (DSIP)

In the oldest modern day study of sleep factors, Monnier and his group in Switzerland have studied the humoral transmission of sleep via the circulatory system. In these studies the jugular vein of a donor rabbit is connected to the jugular vein of a recipient rabbit. Subsequently the donor's mediocentral intralaminary thalamus is stimulated at low frequencies (6/s). Such stimulation induces synchronization of the cortical EEG or what this group calls "orthodox delta sleep" in both the donor and recipient (Monnier et al. 1963). Since the animals are joined by the jugular vein, a sleep substance released into the cerebral venous blood of the donor (as a result of thalamic stimulation) is thought to penetrate the recipient's brain via its heart, lung, and carotid artery, thus showing that sleep can be induced by blood-borne factors. Similar results have been obtained in parabiotic rats (Matsumoto et al. 1972).

Since stimulation of the activiting midbrain reticular formation induces arousal with EEG low voltage fast activity in the recipient rabbit, Monnier's results imply that the "sleep" humoral factor is produced as result of being *awake.* This is at variance with Pappenheimer's results, although the experimental approaches of these two authors are not comparable because of the striking differences in methodology.

In subsequent studies, Monnier and Hosli (1964, 1965) dialyzed the cerebral venous blood and showed that the dialysate from a sleeping donor rabbit induced sleep in recipient rabbits. The assessment of the blood-borne hypnogenic factor was originally made by intravenous injections. Later, testing methods were improved by the use of intraventricular infusions (Monnier and Hatt 1971), and it was shown that a partially purified sleep dialysate induced mild bradycardia but had no effect on respiratory rate and CSF and blood pressures, suggesting that its sleep-inducing properties could not be attributed to visceral factors (Monnier et al. 1973).

The assessment of sleep electrographic data was also improved by expressing an average delta value of 200 mm for a period of 5 min or 2360 μV/min, chose as a baseline equivalent to a 100%. A deflection above 100% indicated a tendency to sleep and below 100% to arousal. Thereafter, all studies expressed sleep as the average delta value in periods of 5 min during 25 min samples, and in percentage deviations from a 100% prior to and following infusions. This method provided a relatively rapid bioassay of "sleep."

With this, they initiated a series of studies leading to the isolation, purification, and physical-chemical characterization of a substance which turned out to be a nonapeptide, which they called *delta sleep-inducing peptide (DSIP)* with a molecular weight of 849 and whose amino acid sequence is Trp-Ala-Gly-Asp-Ala-Ser-Gly-Glu (Monnier et al. 1972, 1977; Schoenenberger et al. 1972, 1977, 1978; Schoenenberger and Monnier 1977). In their more recent experiments, assessment of EEG activity was carried out using linear and three-dimensional power spectra analysis, as well as factor analysis to determine variations of spectral energies in frequency bands of 2 Hz (Monnier et al. 1977; Schoenenberger et al. 1978). The experiments leave little doubt that DSIP induces a higher percentage of slow waves.

In a recent study, Polc et al. (1978) tested the effects of intravenous administration of 30 nmol and 300 nmol DSIP on the sleep-wake cycle of cats recorded by telemetry for 6 h. Both SWS and REM sleep was increased by the low dose of DSIP, but the high dose had no effect. However, recently Walker and Sandman injected 59 nmol DSIP into the lateral ventricle of rats and found no decrease in locomotor activity (if anything they found an increase). In our laboratory 235 nmol DSIP applied to the midbrain reticular formation (MRF) of cats produced no changes in sleep (see Kastin et al. 1978). In the light of Polc's results, such local application of DSIP may represent a very high dose, which could explain the absence of effects on sleep and locomotor activity.

However, more recently Tobler and Borbely (1980) have administered 7, 24, 40, 80, and 160 nmol DSIP to rats and found no decrease in locomotion and no increase in sleep. Again in these studies if anything there was an increase in activity. The authors concluded that DSIP does not qualify as a specific sleep-promoting substance.

In addition, the absence of a dose-effect is quite strange. Schoenenberger et al. (1978) have justified the absence of a dose-effect, writing: "DSIP induces EEG changes similar to those of natural sleep, but differing from the effects of hypnotic pharmaca. The chief requirement for classical hypnotic pharmaca (a linear dose-effect curve) was found unsuitable for the new group of DSIP compound. Indeed, intravenous administration of DSIP to free moving rabbits increases only at a definite optimal dose, with a narrow dose-effect relationship the orthodox delta and spindle sleep (slow wave sleep) the corresponding paradoxical sleep and the decreases in motor activity."

The Monnier group suggests that DSIP acts more as a modulator or programmer of sleep than as a neurotransmitter. Whatever the role of DSIP, we should anticipate studies from this and other groups on the origin, distribution, and receptor locations of this substance in the brain. In view of the conflicting evidence one should carefully evaluate the evidence that DSIP is a true sleep substance.

Moreover, there are a few isolated studies which have provided evidence against the presence of blood-borne sleep factors. Ringle and Herndon (1968) failed to find dialyzable sleep-promoting factors in plasma of sleep-deprived rabbits. However, this study was carried out to determine whether sleep deprivation would promote the appearance of a sleep substance, whereas in Monnier's studies the idea was that sleep itself would promote it. The difference in results may therefore be attributed to a difference in approach. Indeed it may be recalled that Monnier and Hosli (1964) suggested that sustained wakefulness elicited by stimulation produced a "waking factor."

In other studies, it was shown by Lenard and Schulte (1972) that craniopagus twins with a common circulation have independent sleep-wake cycles, and that a dog with an extra head transplanted into its circulation (De Andres et al. 1976) failed to present simultaneous sleep periods. These observations, however, cannot really be taken as a serious criticism of DSIP, because under such abnormal conditions, it is improbable that levels of brain excitability are comparable to those of normal subjects.

In summary, on the basis of the available data, we have to be skeptical about DSIP, essentially because the bulk evidence of its sleep-inducing properties comes from observations on changes in EEG amplitude. The association of a humoral agent with one electrographic sign, does not necessarily mean that sleep is being produced.

3.2.3 Sleep Factors from Brain

Recently there have been attempts to extract sleep-promoting material from the brain. Here again, the approaches have been based on different hypotheses concerning the periods in which this presumed material should be obtained.

Nagasaki et al. (1974), evidently influenced by the Pierón concept that a "sleep substance" accumulates during wakefulness, deprived rats of sleep for a period of 24 h by applying an electric shock whenever the animals attempted to sleep. At the end of the deprivation period the rats (n=1,000) were sacrificed and the brain stem was separated, homogenized, dialyzed, and lyophylized. This crude brain extract was injected intraperitoneally into rats and was reported to induce a decrease in locomotor activity and an increase in delta activity and SWS time. Unfortunately, this assessment was performed in a very small number of animals (n=3 for locomotor activity; n=4 for sleep time). Nonetheless, the experiment suggests that the brain stem contains a sleep-promoting material which accumulates during wakefulness and which appears to cross the blood-brain barrier. In a subsequent study (Nagasaki et al. 1976) this brain extract from sleep-deprived rats (BE-SRD) was partially purified by Sephadex G-10 and thin-layer chromatography. It has a molecular weight of 200–700 and produces inhibition of the spontaneous discharge rates of the crayfish abdominal ganglion. Since this inhibitory effect was 100 times stronger than GABA, while 5-HT increased firing rate, the authors suggest that BE-SDR contains peptide-like substances with strong inhibitory properties. It should be noted that these experiments provide support for Piéron's concept that during wakefulness the CSF accumulates a "hypnotoxin" with inhibitory properties.

Other studies have been performed in our laboratories. The approach was influenced by concepts of synaptic transmission, in the sense that the extraction of a sleep-inducing substance should be made at the time it is released, presumably as a result of a physiologic activation during sleep. In these studies, push-pull cannulae were used (Drucker-Colín and Spanis 1976 for description), which permit the extraction in freely moving animals of "active" substances from specific brain sites. The experiments were based on the simplistic notion that sleep and waking can be explained by a reciprocal interplay of two basically antagonistic neuronal systems, as elegantly described by Moruzzi (1972). Accordingly, experiments were undertaken to extract from the midbrain reticular formation ("waking system") substances released during sleep into the extracellular fluid (by nerve terminals of the "sleep system") and to test the effects of such substances on the homologous MRF neurons of a recipient animal. These experiments showed that the substances (perfusates) obtained from cats during sleep were capable of inducing sleep when perfused into the homologous MRF of recipient cats. Since the assessment was based on recordings of the entire sleep-wake cycle, it was concluded that perfusates obtained from specific brain sites during sleep contained substances capable of influencing ongoing waking behavior (Drucker-Colín et al. 1970; Drucker-Colín 1973). In these types of experiments, isolation of a sleep-inducing factor from perfusates is a serious problem, primarily because extraction though push-pull cannulae yields extremely low quantities of material and because bioassays based on lengthy sleep-wake cycle recordings makes trial and error assessment at various stages of purification an overwhelming task.

3.2.4 REM Sleep Polypeptide Factor

Several years ago, Oswald (1969) suggested that the rebound of REM sleep that occurs during drug withdrawal reflects a phase of neuronal repair indicated by increased protein synthesis and further suggested that situations in which protein synthesis increases should lead to REM sleep. For an excellent restatement of these ideas see Adam and Oswald (1977). Since the discovery of Takahashi et al. (1968) of a relationship between growth hormone (GH) secretion and sleep (stages 3 and 4), several studies have been carried out, further investigating this association (Martin 1976; Quabbe 1977). Although there are some studies which have shown that nocturnal GH secretory bursts may occur in some subjects despite an almost total absence of SWS (Sassin et al. 1969; Karacan et al. 1971), in general it can be said that GH secretion is associated with SWS. It remains to be determined, however, whether the former is part of the neural processes involved in the regulation of sleep or merely part of the output system. A role favoring GH as part of the triggering mechanisms of sleep may stem from the fact that in normal adult mammals one of the more stable characteristics of the sleep cycle is that SWS always precedes the occurrence of REM sleep. Since in man the initial periods of SWS are the ones which produce the pronounced release of GH, the possibility has been suggested (Stern and Morgane 1977) that such a burst of GH release may play a role in triggering subsequent REM episodes. Unfortunately no such

GH release in SWS has been found in mammals. This may be due to the fact that mammals are very polycyclic and that it would thus be difficult to detect GH release in SWS. This is partially supported by Takahashi's et al. (1978) observations that sleep-deprived dogs do show GH release in SWS, which, as he has suggested, may indicate that length of waking is a factor in GH release in SWS. Despite these problems, the experiments to be described below seem to suggest that GH may in fact indirectly be part of a system which is involved in triggering REM sleep, but that this is mediated by protein synthesis. Some evidence bearing on this suggestion has come from experiments which have shown that the administration of GH to rats (Drucker-Colín et al. 1975) and cats (Stern et al. 1975) produces a dose-dependent and specific increase in REM in the 3rd h following its administration. A similar increase in REM sleep has recently been shown to occur following GH administration to humans (Mendelson et al. 1980). It should be pointed out that GH is one of the few substances which have been shown to increase REM time across several species including man.

It would be expected from these results that somatostatin, which is a GH inhibiting factor (Brazeau et al. 1973), would produce a decrease or at least alterations in sleep, since it is well known that its administration postpones GH secretion in humans (Lucke et al. (1976). Some evidence of disturbed sleep has been reported by Havlicek et al. (1976) and Rezek et al. (1976) who observed that in rats small doses of somatostatin reduced SWS and almost completely abolished REM sleep. These studies, however, were obscured by the fact that somatostatin induced motor excitation and prolonged spells of compulsory scratching and circular movements, making it difficult to determine whether this peptide altered some basic sleep mechanisms.

It is well known that GH has important anabolic functions. Moreover, according to Tata (1968) GH seems to increase the rate at which newly formed ribosomes occur and with it the enhancement of protein synthesis. Moreover, such effects are additive when two different anabolic hormones are administered. However, since different hormones seem to synthesize different specific proteins, it is suggested that two anabolic hormones bring about similar effects in the same tissue but through different routes (Tata 1968). This is interesting for sleep in view of the fact that 3 h after GH administration, when REM was found to be increased, there is also an increase in protein levels (Drucker-Colín et al. 1975) and of protein synthesis in the diencephalon and pons medulla (Stern and Morgane 1977), evidently supporting the idea that GH produces its effect on REM sleep by way of its protein synthesizing capabilities. Unfortunately, there are scanty data on the relationship between protein synthesis and sleep. Stern et al. (1972) reported an increase in REM sleep for 7 days after the intraventricular administration of cycloheximide in cats. Since protein synthesis inhibition was at 75% on the 1st day and at 50% by the 4th day, they suggested that the increase in REM sleep was due to protein synthesis patterns in the brain which were returning towards a normal state. In this same study, cycloheximide had no effect on sleep when injected intraperitoneally. However, Pegram et al. (1973) reported that such injections of cycloheximide in mice produced a specific decrease in REM sleep. We have also observed a specific decrease in REM sleep in rats (Rojas-Ramírez et al. 1977) and cats (Drucker-Colín et al. 1979 a) following the administration of protein synthesis inhibitors such as anisomycin and chloramphenicol. Similar observations with administration of chloramphenicol have been made by Kitahama

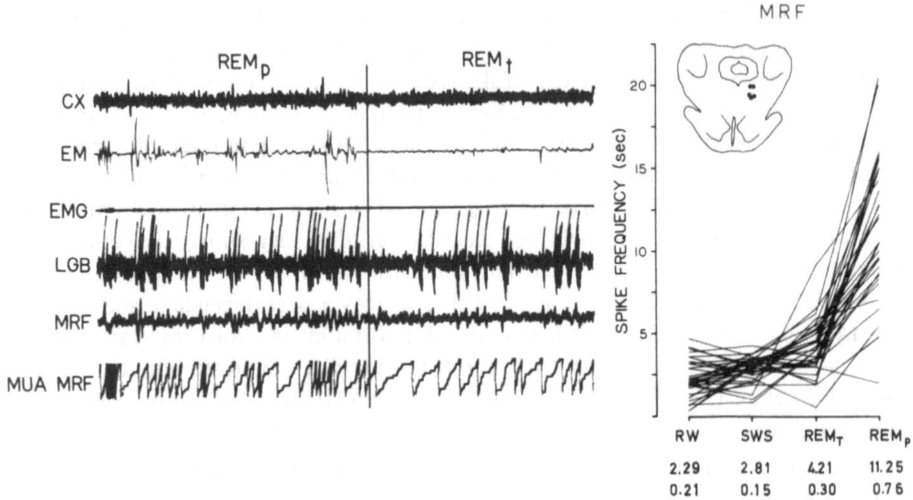

Fig. 2. Samples of REM sleep in which it can be clearly seen that this phase can be divided into a period in which there is a high level of phasic activity *(REM_p)* and one in which such activity has decreased and represents the tonic period *(REM_t)*. Normally the cat in REM sleep constantly oscillates between these two periods. *Cx,* cortex; *EM,* eye movements; *LGB,* lateral geniculate body; *Hip,* hippocampus; *MRF,* midbrain reticular formation; *MUA,* multiple unit activity. On the *right* is a graph which shows that spike frequency in MRF is very high during REM_p, indicating that the cause of the overall increase in spike frequency normally seen during REM sleep as compared to SWS or wakefulness are the events occurring during REM_p. (Modified from Drucker-Colín et al. 1977)

and Valatx (1975) and by Petitjean et al. (1979) in cats. It is important that in all these studies SWS was unaffected except when drugs were administered at very high doses (Petitjean et al. 1979). It is interesting that the decrease in REM is due to a reduction in frequency, since mean duration is unaffected (Pegram et al. 1973: Rojas-Ramírez et al. 1977; Drucker-Colín et al. 1979). These results may suggest that protein synthesis could be involved in the mechanisms that trigger REM sleep.

This probability was tested by determining the effects of chloramphenicol on REM sleep rebound, induced either by REM sleep deprivation or by withdrawal of chronic amphetamine administration. In both cases chloramphenicol specifically blocked the rebound of REM sleep by decreasing its frequency of appearance (Drucker-Colín and Benitez 1977).

In addition experiments have shown that protein synthesis inhibitors also affect some of the components of REM sleep, mainly the phasic events. Observation of any REM sleep period clearly indicates that it is not a homogeneous period. This means that eye movements, myolonic twitches, high frequency bursts of unit acitivity, and PGO spikes constantly oscillate with periods in which these events are absent or diminished (Fig. 2). Although it has often been reported that the frequency of unit activity (which is a good indicator of phasic events) during REM sleep is higher than that seen during any other phase of the sleep-wake cycle (Steriade and Hobson 1976),

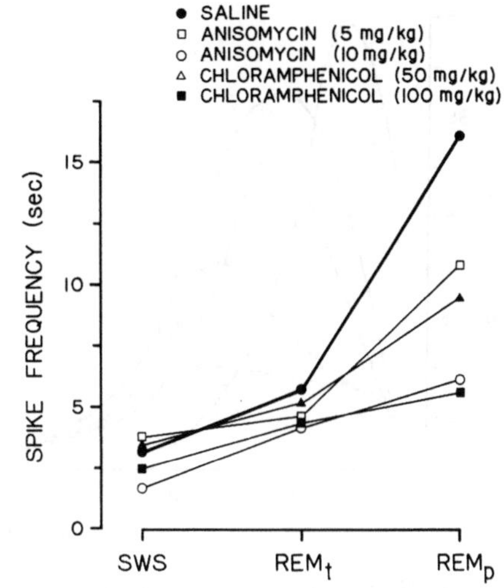

Fig. 3. Mean spike frequency during *SWS*, *REM$_t$*, and *REM$_p$*. Note how protein synthesis inhibitors produce a dose-dependent decrease in spike discharge in MRF during *REM$_p$*. This merely reflects the abolition of bursting periods of spike discharge

such increase is due solely to the bursting periods, since in their absence unit activity is as low as that seen in other stages (Drucker-Colín et al. 1977). For this reason, REM sleep should be analyzed in two distinct periods, phasic REM sleep (REM$_p$) and tonic REM sleep (REM$_t$), as in Fig. 2. It is very interesting to note that protein synthesis inhibitors have a very clear and potent effect on multiple unit activity (MUA) frequency during REM$_p$. In Fig. 3, we can observe that the typical spike bursting period is abolished by chloramphenicol, and that this effect is dose dependent. Figure 3 gives the mean discharge frequencies during some 30 periods of relaxed wakefulness, SWS and tonic and phasic REM. It is clear that only the phasic bursts during REM are affected. We have recently calculated that protein synthesis inhibitors decrease the amount of phasic REM to about 50% (Drucker-Colín et al. 1979a).

Although there is no clear indication of how protein synthesis inhibitors and the decrease of phasic activity affects the normal occurrence of REM sleep, we have recently obtained some evidence suggesting that they may affect both its triggering and its duration. Utilizing MUA from MRF neurons as an indicator of phasic activity, we calculated the discharge frequency in 30 s periods 1 min prior to onset of REM and 1 min during REM. In addition REM periods of over 2 min duration were compared with those of less than 2 min. An interesting observation emerged. When REM sleep periods were longer than 2 min in duration, spike frequency increased just prior to onset of REM and showed a further increase during the 1st min of that state. Similar events have been recorded in pontine reticular structures (Hobson et al. 1974). On the other hand, when REM sleep duration lasted less than 2 min, no such positively inflected rise in discharge frequency was observed (Drucker-Colín et al. 1979b).

Since, as mentioned, it is phasic discharge of unit activity which contributes to the rise in frequency observed during REM sleep, these results suggest that in the absence of phasic discharges, REM sleep will be short or abortive. Most interesting is the observation that in cats given protein synthesis inhibitors the discharge rate of MRF neurons is practically identical to that which occurs in short duration REM sleep

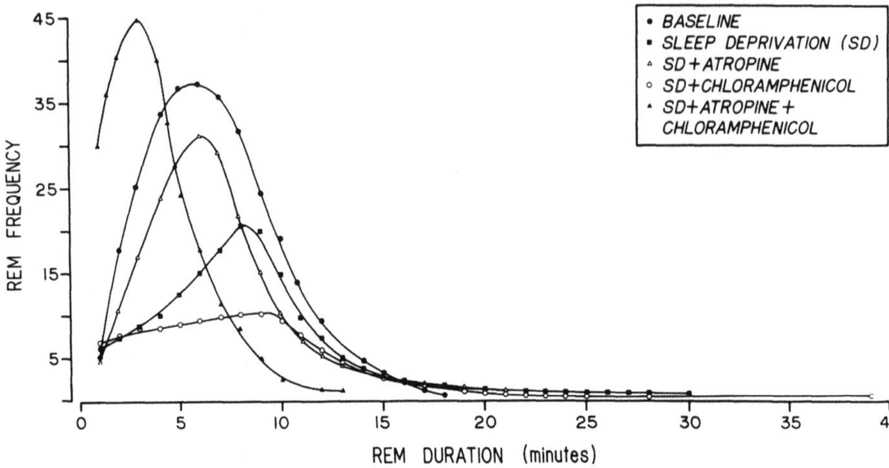

Fig. 4. Represents a plot showing the amount of times a given REM period occurred as a function of its duration. It can be clearly observed that following 72 h of sleep deprivation there is an increase in the duration of the REM sleep periods, so this curve shifts towards the right. On the other hand, if atropine and chloramphenicol are administered following 72 h of sleep deprivation there are an inordinate number of very short REM periods (for further details see text)

periods (Drucker-Colín et al. 1979b). This could indicate that by preventing phasic activity, these drugs reduce the probability of REM sleep periods occurring, and shorten them when they occur, because of the absence of phasic activity. Unfortunately, the absence of phasic activities with protein synthesis inhibitors is not always accompanied by an absence of long REM sleep periods. In fact, these drugs do not alter REM sleep duration (Drucker-Colín et al. 1979a). This may be related to the fact that protein synthesis inhibitors, while decreasing MUA bursts, do not alter PGO spike frequency, which represents another type of important phasic event. Since atropine can diminish PGO spike density (Hendriksen et al. 1972), we carried out some experiments in cats deprived of sleep for 72 h in which we administered either atropine or chloramphenicol or both. We observed that both atropine and chloramphenicol reduced the REM sleep rebound, but while the latter did so through REM frequency, the former did it through REM duration. But most interesting was that the simultaneous administration of these drugs induced an inordinate number of very short REM sleep periods. It must be recalled that these animals had 72 h without REM sleep. Figure 4 illustrates the dramatic change in REM length following the combined drug treatment. Here we graphed the frequency of REM periods according to their duration. It can be clearly seen that atropine + chloramphenicol shifts the mean REM duration towards the left, and that this is caused by a great number of REMs of 1 and 2 min duration. The main characteristic of REMs under atropine + chloramphenicol is that of a very low level of phasic activity. We therefore suggest that under normal conditions REM duration may depend on a certain level of neuronal bombardment. When this falls below a minimal level, as in the cats treated with atropine and chloramphenicol, the animals wake up shortly after the onset of REM sleep.

It seems evident that both cholinergic and polypeptidergic mechanisms participate in some fundamental manner in REM sleep. This has become even more evident in experiments using the model of REM sleep without atonia. This model was described for the first time in 1965 (see Jouvet 1972), and has been studied more recently and more extensively by Morrison (1979) and Sastre and Jouvet (1979). In these cats small bilateral lesions of the locus ceruleus *a* (LC) and its vicinity do not prevent the animal from going into REM sleep, but do prevent the muscular atonia characteristic of this state. Moreover, such preparations during REM sleep present several stereotyped patterns of behavior such as licking, grooming, lunging attacklike movements, and even standing and walking.

Magon and Rhines (1946) described an inhibitory bulbar reticulospinal pathway whose cells of origin have recently been shown to be postsynaptically excited by LC*a* neurons (Kanamori et al. to be published). This area has also been shown to receive direct afferent projections from LC*a* (Sakai et al. 1979). The atonia of REM sleep following LC*a* lesions could then be explained by the absence of excitatory inputs into the bulbar reticulospinal inhibitory pathway of Magoun and Rhines, as can be seen in Fig. 5a. The movements observed in LC*a* lesioned cats could then be due to the occurrence of phasic bursts of unit activity impinging on the motor cortex, which in turn activate pyramidal tract neurons back unto motoneurons which are no longer inhibited by the reticulospinal pathway (Fig. 5b). Sastre (1978), however, suggests that the movements during REM sleep are caused by PGO spikes, because the periods of loss of muscle tone in LC*a* lesioned cats are accompanied by a decrease in PGO spikes.

The question of whether PGO spikes or reticular phasic activities are mainly responsible for the movements in LC*a* lesioned cats can easily be tested since it is easy to dissociate phasic activities by pharmacologic means, i.e., atropine decreases the number of PGO spikes and chloramphenicol decreases MUA bursts. We have thus lesioned the LC*a* in a few cats and obtained a series of motor activities during REM sleep. We then administered either 0.8 mg/kg atropine or 75 mg/kg chloramphenicol. We observed that a decrease in PGO density altered neither the intensity nor the frequency of motor behavior during REM sleep. On the other hand, chloramphenicol, while completely eliminating the bursts of MUA, also produced a disappearance of the motor patterns of behavior and significantly increased the time spent in muscle atonia (Fig. 6).

Since it has recently been shown by Ramirez (1973) that chloramphenicol produces a strong inhibition of synaptic plasma membrane proteins, it is conceivable that such proteins play an important role in regulating sleep, perhaps through their effects on certain types of phasic activities. The evidence so far, however, has been provided by indirect neuropharmacological experiments. We have however, more direct experimental evidence. In one such series of experiments cats implanted with electrodes for recording the sleep-wake cycle and a push-pull cannula in the MRF were continuously perfused and recorded for 12–21 h. At every hour the protein content of the perfusate and the time spent in wakefulness, SWS and REM sleep was determined. It was found that protein concentrations in the perfusate varied in a cyclic fashion, and that peaks of protein concentration corresponded to those hours in which REM sleep occupied the greatest proportion of time (Drucker-Colín et al. 1975). Moreover, when insomnia

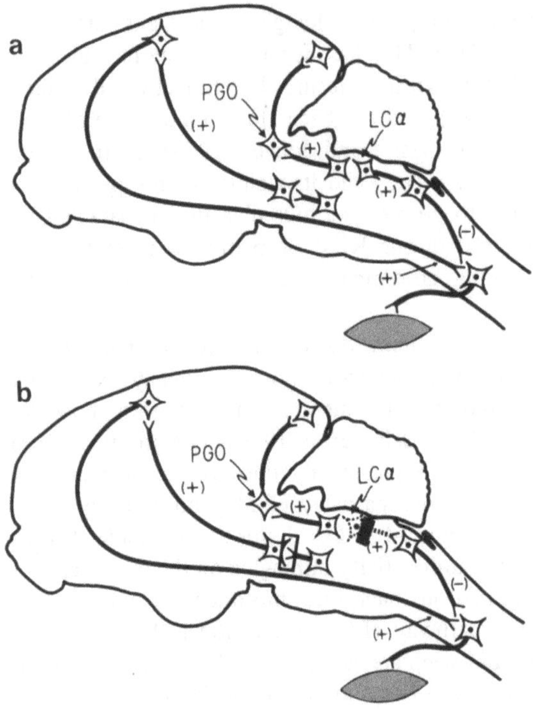

Fig. 5 a and b. Schematic model of events occurring during REM sleep. **a** Normal cat where muscle atonia would be seen as arising from excitation of locus coeruleus *α (LC)* cells impinging upon the bulbar reticulospinal inhibitory pathway. The PGO spikes would arise from a generator located in the pons (see text), and the phasic unit activity would arise from generator neurons in the pons which phasically excite MRF neurons which in turn excite cortical neurons in motor cortex. These cells activate the pyramidal tract, which is excitatory upon the motoneurons, but these influences normally cannot override the powerful reticulospinal inhibition. The myoclonic twitches occasionally seen during REM sleep could be caused by a momentary loss of such inhibition. **b** When LC*α* neurons are lesioned *(black box)*, the excitation of the inhibitory pathway is no longer present, and as a consequence the pyramidal tract excitation predominates and REM sleep without atonia ensues. Since in general these animals present phasic bursts of behavior, it is suggested *(empty box)* that when chloramphenicol is administered, which eliminates the phasic bursts of unit activity, the cat should present a significant loss of movements even in the face of LC *α* lesions. In these experiments PGO spikes are seen as playing little or no role in behavior during REM sleep without atonia

was produced by bilateral lesions in the preoptic area, the cyclic release of proteins disappeared (Drucker-Colín and Gutiérrez 1976). Protein levels also lost their cyclicity following the administration of chloramphenicol, which specifically decreases REM sleep. Moreover when waking perfusates were compared to those from REM sleep, the latter always contained nearly twice as much protein (Drucker-Colín and Spanis 1975). Similar observations when compared to SWS have been made by Brodskii et al. (1974). In further studies, separation of released proteins by polyacrylamide SDS gel

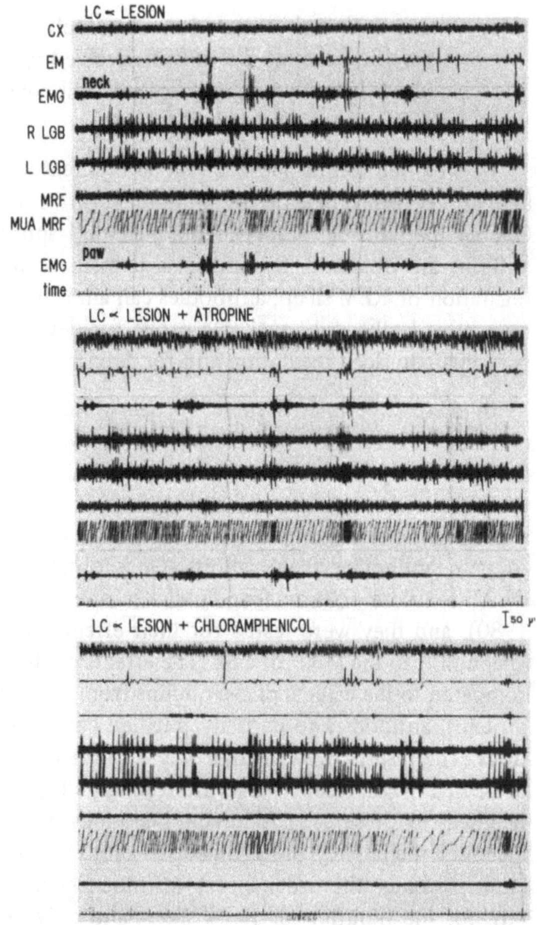

Fig. 6. Electrophysiological recordings in a cat with LC *α* lesion during REM sleep. The *upper graph* shows the presence of muscle tone and bursts of movement, some quite ample. When this animal is given 0.8 mg/kg atropine, the PGO spike density diminishes greatly, but there is no effect on muscle tone. However, when the animal is administered 75 mg/kg chloramphenicol, which eliminates the bursts of MUA, then the muscle tone decreases and there is absence of movement (see text for further explanation)

electrophoresis revealed the existence of some large molecular weight proteins in REM sleep perfusates which are not present in waking perfusates (Spanis et al. 1976). However, recently we have been unable to confirm this finding. The differences between the two states, therefore, seem so far to be merely that of concentration, although the possibility that specificity of the proteins could be detected with more sensitive methods should not be ignored.

Although these experiments are highly suggestive that proteins may be involved in the regulation of REM sleep, the evidence is merely correlative. Moreover, it is impossible to determine whether the release of the proteins is the cause or the effect of REM sleep. We have therefore tested an approach which in a more direct form explores whether specific proteins are indeed involved in the regulation of REM sleep. This approach consists in studying the effects of antibodies against proteins which are obtained from the MRF of cats during REM sleep.

We were led to this approach for several reasons. Firstly, antibodies can be obtained in relatively large quantities by immunization with small amounts of perfusate proteins. One of the limiting factors of the push-pull approach is that insufficient amounts of antigens are obtained. Second, if a protein or polypeptide is indeed involved in the regulation of REM sleep, antibodies can amplify the effects on sleep. Third, antibodies may provide the only available probe for determining whether such molecules really participate in sleep regulation and for determining receptor regions.

In these experiments some 40 cats were implanted for recording the sleep-wake cycle and with a push-pull cannula in the MRF. Collection of perfusate proteins was first done only when the animals were in REM sleep. However, since it was later found that this was not necessary (Drucker-Colín et al. to be published), collection of material was carried out continuously. The proteins obtained from the perfusate were utilized to raise antibodies from rabbits. Once obtained, the antibodies were purified on a DEAE-cellulose column (for details of the entire procedure see Drucker-Colín et al. 1980), and they were tested for their effects on the sleep-wake cycle following their injection into the MRF of cats. The effects of these antibodies anti-MRF proteins were compared to the effects of α-globulins from preimmune serum, antibodies to cat serum proteins, antibodies to partially purified cat albumin, Ringer solution, and sham injections. The results showed that only anti-MRF antibodies significantly and specifically decreased REM sleep. Most interestingly, these antibodies produced effects almost identical to those seen following injections of chloramphenicol. First, the antibodies increased the latency of REM sleep to about 5 h, in contrast to all controls whose mean latency was in the order of 50–60 min. Second, antibodies decreased the frequency but not the duration of REM sleep, and only at high doses (500–1,000 µg) did the antibodies have any effect on SWS. This effect on SWS, which manifested itself as an increase in the latency, was only transient since total SWS time was unchanged in the 12 h recorded. In addition to the effects on REM sleep, the anti-MRF antibodies also decreased the MUA bursts, but had no effect on PGO spikes (Drucker-Colín et al. to be published).

The effects of these antibodies provide the most direct evidence for a role of macromolecules in the regulation of REM sleep. However, a word of caution should be advanced. First, it should be emphasized that these are preliminary experiments and that the antibodies are a mixture of several. Second, since antibodies raised against proteins obtained during wakefulness or proteins from another area of the brain have similar effects, it is possible that we are dealing with nonspecific effects. However, since other antibodies failed to alter REM sleep, nonspecificity does not seem to be responsible for the results. It is thus plausible that a specific polypeptide is always present in the extracellular fluid and is only activated during REM or reaches adequate levels at that particular time. Some support for the latter comes from the fact that during REM sleep there is always at least twice the amount of protein than during wakefulness (Drucker-Colín and Spanis 1975). It is possible, therefore, that the occurrence of REM sleep depends on a "gating" signal from a polypeptide, which then transduces its signal into a specific type of firing pattern, which in turn stimulates a series of structures which put out the various neuronal activities producing REM sleep as it is normally seen, for example, atonia, eye movements, PGO waves, and phasic unit activity. The latter two in particular seem additionally to provide the nervous system during REM sleep with a failsafe mechanism to preserve it and to insure its adequate duration. If both activities fall below a certain level of intensity, REM sleep can no longer sustain itself.

4 The New Neurobiology of Sleep

Under this rather pretentious title we wish to make some concluding remarks. Such remarks will evidently not solve the problem of the humoral regulation of sleep, but may draw the reader into thinking in a broader fashion about the sleep-wake cycle. First, it is evident that there is no single sleep substance, aside from the fact that such substances has not yet been discovered, and second, that the brain is by itself a target organ for its own neuroendocrine outputs.

In view of this it seems reasonable to postulate that the phase relationship among constituent rhythms within the circadian system is an essential and critical aspect for the occurrence and timing of the major part of the sleep-related neuroendocrine events that characterize this portion of the cycle. In this connection the role played by hormones and peptides seems to be of paramount importance. They represent a group of highly specific signals acting throughout, triggering changes in the expression of the genomic and/or membrane information and producing differences (both qualitatively and quantitatively) in the biosynthetic machinery of practically all the cells in the organism (for review see Baxter and Funder 1979). In terms of sleep this would agree with the ideas recently put forth by Jouvet (1978), in which he suggests that REM sleep may be genetically programmed. However, although the genomic expression may be involved in triggering REM sleep, it is evident that the multifactorial elements of this phase and their involvement in its generation are barely known. It goes without saying that SWS presents the same problems.

It is therefore very important to change our manner of interpreting the view in which the onset of sleep occurs. The one factor, one stage mechanism of sleep is no longer functional, since it is based on classic synaptology and does not take into account the existence of pacemakers made up of various coupled oscillators and the possible existence of ultralow frequency neuronal cycles (Enright 1980), which may program or set in motion the various elements which physiologically we come to recognize as sleep. Such oscillators are probably well designed to be influenced by hormone or peptidelike substances. Despite the complexity of these molecules, sleep research will greatly benefit if, in the future, attempts are made to incorporate into its mechanism the consideration that they play a prominent role.

One may benefit from thinking that the sleep-wake cycle, including REM sleep, comes about in a manner closely paralleling the models of oscillating systems (Moore 1979). In terms of sleep, then, one may suggest the existence of a principal driving oscillator which passively sets in motion the remaining functions. In this case these secondary functions cannot occur unless the main oscillator is set in motion. This would necessitate the existence of oscillating sleep specific groups of cells. Such cells have yet to be described. Another possibility is that of a series of oscillators arranged in a hierarchical organization. The secondary and tertiary oscillators would enter in phase with the main driving oscillator. Removal of this driving oscillator would not eliminate the entire function, but merely dissociate them. This is somewhat reminiscent of the effects of chloramphenicol and atropine on the various elements of REM sleep as described above (also see Drucker-Colín et al. 1979a). A final possibility is the existence of various oscillating systems which interact, but in a nonhierarchical manner. This would necessitate the existence of various oscillators driven by both internal and external inputs which would couple them to the point of synchronization. There are no examples of this as yet in sleep research. The latter two possibilities are probably more congruent with the mechanisms by which sleep comes about and are more consistent with the role that hormones and peptides may play within the assembly.

References

Abe K, Kroning J, Greer MA, Critchlow V (1979) Effects of destruction of the suprachiasmatic nuclei on the circadian rhythms in plasma corticosterone, body temperature, feeding and plasma thyrotropin. Neuroendocrinology 29:119–131

Adam, K, Oswald I (1977) Sleep is for tissue restoration. J Coll Physicians Lond 11: 376–388

Amico JA, Stolc LT, Seif SM (1980) Mid-cycle elevation of oxytocin and oxytocin-neurophysin in women (Abstr 346). The Endocrine Society 62nd Annual Meeting. Endocrine Soc, Washington

Annunziato L (1979) Regulation of the tuberoinfundibular and nigrostriatal system. Evidence for different kinds of dopaminergic neurons in the brain. Neuroendocrinology 29:66–76

Antunes JL, Carmel PW, Zimmerman EA (1977) Projections from the paraventricular nucleus to the zona externa of the median eminence of the rhesus monkey: an immunohistochemical study. Brain Res 137:1–10

Aschoff J (1979) Circadian rhythms: general features and endocrinological aspects. In: Krieger DT (ed) Endocrine rhythms. Raven, New York, pp 1–62

Aserinsky E, Kleitman N (1953) Regularly occurring periods of eye motility, and concomitant phenomena during sleep. Science 118:273–274

Barnea A, Oliver C, Porter JC (1977) Subcellular compartmentalization by hypothalamic peptides: characteristics and ontogeny. Adv Exp Med Biol 87:49–75

Baxter JD, Funder JW (1979) Hormone receptors. N Engl J Med 301:1149–1161

Baxter JD, Eberhardt NL, Apriletti JW, Johnson LK, Ivarie RD, Schachter BS, Morris JA, Seeburg PH, Goodman HM, Latham KR, Polansky JR, Martial JA (1979) Thyroid hormone receptors and responses. Recent Prog Horm Res 35:95–153

Beaumont A, Hughes J (1979) Biology of opiod peptides. Annu Rev Pharmacol Toxicol 19:245–267

Ben-Jonathan N, Oliver CH, Weiner HJ, Mical RS, Porter JC (1977) Dopamine in hypophysial portal plasma of the rat during the estrous cycle and throughout pregnancy. Endocrinology 100:542

Bentley PJ (1976) Comparative vertebrate endocrinology. Cambridge University Press, Cambridge Melbourne New York

Bergland RM, Page RB (1980) Pituitary-brain vascular relations: a new paradigm. Wislocki's models for brain-pituitary relations emphasizing portal "veins" is reconsidered and revised. Science 204:18–29

Bloom FE (to be published) Peptides: integrators of cell and tissue function. Soc Gen Physiol Ser 35

Bowie EP, Herbert DC (1976) Immunocytochemical evidence for the presence of arginine vasotocin in rat pineal gland. Nature 261:66

Boyar RM (1978a) Control of the onset of puberty. Annu Rev Med 29:509–528

Boyar RM (1978b) Sleep related endocrine rhythms. In: Reichlin S, Baldessarini RJ, Martin JB The hypothalamus. Raven, New York, pp 373:386

Brazeau P, Vale W, Burgus R, Ling N, Blutcher M, Rivier J, Guillemin R (1973) Hypothalamic peptide that inhibits the secretion of immunoreactive pituitary growth hormone. Science 179:77–79

Brodskii V, Gusatinskii NV, Kogan AB, Mechaeva NV (1974) Variations in the intensity of ^{3}H-leucine incorporation into proteins during slow wave and paradoxical phases of natural sleep in the cat associative cortex. Dokl Akad Nauk SSSR 215: 748–750

Brownfield MS, Kozlowsli GP (1977) The hypothalamo-choroidal tract I. Immunohistochemical demonstration of neurophysin pathways to telencephalic choroidal plexuses and cerebrospinal fluid. Cell Tissue Res 178:111–127

Brownstein MJ, Russel JT, Gainer H (1980) Synthesis, transport and release of posterior pituitary hormones. Science 207:373–378

Candland DK, Horowitz SH, Culbertson JL (1962) Acquisition and retention of acquired avoidance with gentling as reinforcement. J Comp Physiol Psychol 61: 50–58

Chihara K, Arimura A, Schally AV (1979) Effect of intraventricular injection of dopamine, norepinephrine, acetylcholine and 5-hydroxitryptamine on immunoreactive somatostatin release into rat hypophyseal portal blood. Endocrinology 104:1656–1662

Cicero TJ (1980) Effects of exogenous and endogenous opiates on the hypothalamic-pituitary-gonadal axis in the male. Fed Proc 39:2551–2554

Cohen DH, Cabot JB (1979) Toward a cardiovascular neurobiology. TINS 2:273–276

DeAndres I, Gutierrez-Rivas E, Nava E, Reinoso-Suarez F (1976) Independence of sleep-wakefulness cycle in an implanted head "encephale isolé". Neurosci Lett 2:13–18

Defendi R, Zimmermann EA (1978) The magnocellular neurosecretory system of the mammalian hypothalamus. In: Reichlin S, Baldessarini RJ, Martin JB (eds) The hypothalamus. Raven, New York, pp 137–152

DeWeid D, Versteeg DH (1979) Neurohypolyseal principles and memory. Fed Proc 38:2348–2354

Dogterom J, Pevet P, Buijs RM, Snijdewint SM, Swaab DJ (1979) Vasopressin, oxytocin and vasotocin in pineal gland, subcommisural organ and foetal pituitary: failure to demonstrate vasotocin in mammals. Acta Endocrinol [Suppl] (Copenh) 225:413

Drucker-Colin RR (1973) Crossed perfusion of a sleep inducing brain tissue substance in conscious cats. Brain Res 56:123–134

Drucker-Colin RR, Benitez J (1977) REM sleep rebound during withdrawal from chronic amphetamine administration is blocked by chloramphenicol. Neurosci Lett 6:267–271

Drucker-Colin RR, Gutiérrez MC (1976) Effects of forebrain lesions on release of proteins from the midbrain reticular formation during the sleep-wake cycle. Exp Neurol 52:339–344

Drucker-Colin RR, Spanis CW (1975) Neurohumoral correlates of sleep: increase of proteins during rapid eye movement sleep. Experientia 31:551–552

Drucker-Colin RR, Spanis CW (1976) Is there a sleep transmitter? Prog Neurobiol 6:1–22

Drucker-Colin RR, Rojas-Ramirez JA, Vera-Trueba J, Monroy-Ayala G, Hernández Peon R (1970) Effect of crossed-perfusion of the midbrain reticular formation upon sleep. Brain Res 23:269–273

Drucker-Colin RR, Spanis CW, Hunyadi J, Sassin JF, McGaugh JL (1975a) Growth hormone effects on sleep and wakefulness in the rat. Neuroendocrinology 18: 1–8

Drucker-Colin RR, Spanis CW, Cotman CW, McGaugh JL (1975b) Changes in protein in perfusates of freely moving cats: relation to behavioral state. Science 187: 963–965

Drucker-Colin RR, Bernal-Pedraza JG, Diaz-Mitoma F, Zamora-Quezada J (1977) Oscillatory changes in multiple unit activity during rapid eye movement sleep. Exp Neurol 57:331–341

Drucker-Colin RR, Zamora J, Bernal-Pedraza J, Sosa B (1979a) Modification of REM sleep and associated phasic activities by protein synthesis inhibitors. Exp Neurol 63:458–467

Drucker-Colin RR, Dreyfus-Cortes G, Bernal-Pedraza JG (1979b) Differences in multiple unit activity discharge frequency during short and long REM sleep periods: effects of protein synthesis inhibition. Behav Neural Biol 26:123–127

Drucker-Colin RR, Tuena de Gomez-Puyou M, Gutierrez MC, Dreyfus-Cortes G (1980) Immunological approach to the study of neurohumoral sleep factors: effects on REM sleep of antibodies to brain stem proteins. Exp Neurol 69:563–575

Dupont A, Barden N, Cusan L, Merand Y, Labrie F, Veudry H (1980) β-endorphin and Met-enkephalins: their distribution, modulation by estrogens and haloperidol, and role in neuroendocrine control. Fed Proc 39:2544–2550

Elde R, Hökfelt T (1978) Distribution of hypothalamic hormones and other peptides in the brain. In: Ganong WF, Martini L (eds) Frontiers in neuroendocrinology, vol 5. Raven, New York, pp 1–33

Elde R, Hökfelt T (1979) Localization of hypophysiotropic peptides and other biologically active peptides within the brain. Annu Rev Physiol 41:587–602

Emanuele N, Kirsteins L, Lawrence AM (1979) Brain LH: localization, response to hypophysectomy and ovariectomy. Clin Res 27:250

Enright JT (1980) The timing of sleep and wakefulness. Springer, Berlin Heidelberg New York

Epstein AN (1978) The neuroendocrinology of thirst and salt peptides. In: Ganong WF, Martini L (eds) Frontiers in neuroendocrinology, vol 5. Raven, New York, pp 101–134

Epstein AN (1980) Angiotensin induced water and salt intake. In: Valverde-R C, Aréchiga H (eds) Comparative aspects of neuroendocrine control and behavior. Karger, Basel Paris London New York, pp 104–119

Fencl V, Koski G, Pappenheimer JR (1971) Factors in cerebrospinal fluid from goats that affect sleep and activity in rats. J Physiol (Lond) 216:565–589

Fisher LA, Spinderl ER, Fernstrom JD (1980) Nonapeptide content of the bovine pineal gland (Abstr 11). The Endocrine Society 62nd Annual Meeting. Endocrine Soc, Washington, p 103

Fitzsimons JT (1980) Angiotension in the control of water and sodium intake. In: Bloom FE (ed) Peptides: integrators of cell and tissue functions. Raven, New York, pp 99–107

Flerko B (1980) The hypophysial portal circulation today. Neuroendocrinology 30: 56–63

Frantz AG (1979) Rhythms in prolactin secretion. In: Krieger DT (ed) Endocrine rhythms. Raven, New York, pp 175–186

Fuller RW, Wong DT (1977) Inhibition of serotonin reuptake. Fed Proc 36:2154–2158

Fuxe K, Hökfelt T, Anderson K, Ferland L, Johansson O, Ganten D, Eneroth P, Gustafsson JA, Skett P, Said SI, Mutt V (1978a) The transmitters of the hypothalamus. In: Cox B, Morris ID, Weston AH (eds) Pharmacology of the hypothalamus. University Park Press, Baltimore, pp 31–61

Fuxe K, Ogren SO, Agnati L, Jonsson G (1978b) Further evidence that methergoline is a central 5-hydroxytryptamine receptor blocking agent. Neurosci Lett 9:195–200

Fuxe K, Anderson K, Hökfelt T, Mutt V, Ferland L, Agnati LF, Ganten D, Said S, Eneroth P, Gustafsson JA (1979) Localization and possible function of peptidergic neurons and their interactions with central catecholamine neurons, and the central actions of gut hormones. Fed Proc 38:2333–2340

Gainer H (ed) (1977) Peptides in neurobiology. Plenum, New York London

Gallo RV (1980) Neuroendocrine regulation of pulsatile luteirizing hormone release in the rat. Neuroendocrinology 30:122–131

Georg CPL, Messerli FH, Genest J, Nowaczynski W, Boucher R, Kuchel O, Ortega RO (1975) Diurnal variation of plasma vasopressin in man. J Clin Endocrinol Metab 41:332–338

Gillin JC, Mendelson WB, Sitaram N, Wyatt RJ (1978) The neuropharmacology of sleep and wakefulness. Ann Rev Pharmacol Toxicol 18:563–579

Golde DW (1980) Growth hormone. Ann Intern Med 92:650–662

Grave GD (1977) (ed) Thyroid hormones and brain development. Raven, New York

Grossman MI (1979) Neural and hormonal regulation of gastrointestinal function: an overview. Ann Rev Physiol 41:27–33

Guansing A, Hagen TC, Hojuat S, Lawrence AM (1977) Brain TSH: extra-pituitary localization of immuno and bioassayable TSH-like activity. Clin Res 25:605A

Gudelsky GA, Porter JC (1980) Release of dopamine from tubero infundibular neurons into pituitary stalk blood after prolactin or haloperidal administration. Endocrinology 106:526–529

Guillemin R (1978) Biochemical and physiological correlates of hypothalamic peptides. The new endocrinology of the neuron. In: Reichlin S, Baldessarini RJ, Martin JB (eds) The hypothalamus. Raven, New York, p 155

Guy J, Leclerc R, Vaudry H, Pelletier G (1980) Identification of a second category of MSH-containing neurons in the rat brain (Abstr 117). The Endocrine Society 62nd Annual Meeting. Endocrine Soc, Washington, p 104

Havlicek V, Resek M, Friesen H (1976) Somatostatin and thyrotropin releasing hormone: central effect on sleep and motor system. Pharmacol Biochem Behav 4: 455–459

Hendriksen SJ, Jacobs BJ, Dement WC (1972) Dependence of REM sleep PGO spikes on cholinergic mechanisms. Brain Res 48:412–416

Heritage AS, Stumpf WE, Sar M, Grant LD (1980) Brainstem catecholamine neurons are target sites for sex steroid hormones. Science 207:1377–1379

Hery F, Pujol JF, Lopez M, Macon J, Glowinski J (1970) Increased synthesis and utilization of serotonin in the central nervous system of the rat during paradoxical sleep deprivation. Brain Res 21:391–403

Hexum TD, Hanbauer I, Yang H-YT, Costa E (1980) Secretion of enkephalinlike peptides from dog adrenal gland (Abstr 1799). Fed Proc 39:605

Hobson JA, McCarley RW, Freedman R, Pivik RT (1974) Time course of discharge rate changes by cat pontine brain stem neurons during sleep cycle. J Neurophysiol 37:1297–1309

Hökfelt T, Johansson O, Ljungdahl A, Lundberg J, Schultzberg M, Fuxe K, Goldstein M, Steinbusch H, Verhofstad A, Elde R (1978a) Neurotransmitters and neuropeptides: distribution patterns and cellular localization as revealed by immunocytochemistry. In: Fuxe K, Hökfelt T, Luft R (eds) Central regulation of the endocrine system. Plenum, New York London, pp 31–48

Hökfelt T, Elde R, Fuxe K, Johansson O, Ljungdahl A, Goldstein M, Luft R, Nilsson G, Said S, Fraser H, Jeffcoate SL, White N, Ganten D, Rehfeld J (1978 b) Aminergic and peptidergic pathways in the nervous system with special reference to the hypothalamus. In: Reichlin S, Baldessarini RJ, Martin JB (eds) The hypothalamus. Raven, New York, pp 69–135

Ibata Y, Watanabe K, Kinoshita H, Kubo S, Sano Y (1979) The location of LHRH neurons in the rat hypothalamus and their pathways to the median eminence. Cell Tissue Res 198:381–395

Inouye SIT, Kawamura H (1979) Persistence of circadian rhythmicity in a mammalian hypothalamic "island" containing the suprachiasmatic nucleus. Proc Natl Acad Sci USA 76:5962–5966

Jackson IMD (1978) Extrahypothalamic and phylogenetic distribution of hypothalamic peptides. In: Reichlin S, Baldessarini RJ, Martin JB (eds) The hypothalamus. Raven, New York, pp 217–231

Jackson IMD (1980) Distribution and evolutionary significance of the hypophysiotropic hormones of the hypothalamus. In: Valverde-R C, Arechiga H (eds) Comparative aspects of neuroendocrine control of behavior. Karger, Basel Paris London New York, pp 36–69

Johnston CA, Demarest KT, More KE (1980) Involvement of protein synthesis in the delayed activation of dopamine synthesis in tuberoinfundibular nerves by prolactin. Fed Proc 39:555

Jouvet M (1972) The role of monoamines and acetylcholine in the regulation of the sleep-waking cycle. Ergb Physiol Biol Chem Exp Pharmakol 64:166–307

Jouvet M (1978) Does a genetic programming of the brain occur during paradoxical sleep. In: Buser H, Rougeul-Buser M (eds) Cerebral correlates of conscious experience. Elsevier, Amsterdam, pp 245–261

Jurado JL, Rhodes JM, Fernández-Guardiola A, Valverde R-C (1979) Positive results in two cases of narcolepsy treated with thyrotropin-releasing hormone (TRH) (Abstr 2359). Society Neuroscience 9th Annual Meeting. Soc for Neurosci, Atlanta, Georgia, p 696

Kanamori N, Sakai K, Jouvet M (to be published) Neuronal activity specific to paradoxical sleep in the ventromedial medullary reticular formation of unrestrained cats. Brain Res

Karacan I, Rosenbloom AL, Williams RL, Finley WW, Nursch CJ (1971) Slow wave sleep deprivation in relation to plasma growth hormone concentration. Behav Neuropsychiatry 2:11–14

Kastin AJ, Nissen C, Schally AV, Coy DH (1978) Radioimmunoassay of DSIP-like material in rat brain. Brain Res Bull 3:691–695

Khachaturian H, Sladek JR Jr (1980) Simultaneous monoamine histofluorescence and neuropeptide immunocytochemistry. III. Ontogeny of catecholamine varicosities and neurophysin neurons in the rat supraoptic and paraventricular nuclei. Peptides 1:77–95

King JA (1979) Phylogenetic and anatomical distribution of somatostatin in vertebrates. Endocrinology 105:1322–1329

Kitahama K, Valatx JL (1975) Effet du chloramphenicol et du thiamphenicol sur le sommeil de la souris. Comptes rendues Soc Biol (Paris) 169:1522–1525

Knigge KM, Joseph SA, Hoffman GE (1978) Organization of LRF and SRIF neurons in the endocrine hypothalamus. In: Reichlin S, Baldessarini RJ, Martin JB (eds) Hypothalamus. Raven, New York, pp 49–68

Kozlowski GA, Brownfield MS, Hostetter G (1978) Neurosecretory supply of extrahypothalamic structures: choroid plexus, circunventricular organs and limbic system. In: Bargman W, Oksche A, Polenov A, Scharrer B (eds) Neurosecretion and neuroendocrine activity. Springer, Berlin Heidelberg New York, pp 217–227

Kozlowski GP, Chu L, Hostetter G, Kerdelhue B (1980) Cellular characteristics of immunolabeled luteinizing hormone releasing hormone (LHRH)-neurons. Peptides 1:37–46

Kroll FW (1933) Über das Vorkommen von übertragbaren schlaferzeugenden Stoffen im Hirn schlafender Tiere. Z Gesamte Neurol Psychiatr 146:208–218

Krueger JM, Pappenheimer JR, Karnovsky ML (1978) Sleep promoting factor S: purification and properties. Proc Natl Acad Sci USA 75:5235–5238

Krulich L (1979) Central neurotransmitters and the secretion of prolactin, GH, LH, and TSH. Annu Rev Physiol 41:603–615

LeDourain N (1974) Cell recognition based on natural morphological nuclear makers. Med Biol 52:281

LeDourain N, Smith J, Teillet M-A, Le Lievre C, Ziller C (1980) The natural crest and its developmental analysis in avian embryo chimaeras. TINS 3:39–42

Legendre R, Pieron H (1910) Le probleme des facteurs du sommeil. Resultats d'injections vasculaires et intra-cerebrales des liquides insomniques. Crit Rev Soc Biol (Paris) 68:1077–1078

Legendre R, Pieron H (1911) Du developpement au cours de l'insomie expérimental, des propriétes hypnotoxiques des humeurs en relation avec le besoin croissant de sommeil. Comptes Rendues Soc Biol (Paris) 70:190–192

Legendre R, Pieron H (1912) De la propriéte hypnotoxique des humerus dévelopeés au cours d'une veille prolongée. Comptes Rendues Soc Biol 72:210–212

Leibson L (1979) Endocrinology evolution and evolutionary endocrinology. Perspect Biol Med 23:25–43

Lenard HG, Schulte FJ (1972) Polygraphic sleep study in cramopagus twins. J Neurol Neurosurg Psychiatry 35:756–760

Leowy DD, McKellar S (1980) The neuroanatomical basis of central cardiovascular control. Fed Proc 39:2495–2503

Lewis RV, Kimura S, Stern A, Rossier J, Stein A, Udenfriend S (1980) Enkephalin precursors in bovine adrenal granules. Fed Proc 39:385 A 613

Liotta AS, Gilderslleve D, Brownstein MJ, Krieger DT (1979) Biosynthesis *in vitro* of immunoreactive 31,000-dalton corticitropin/-endorphinlike material by bovine hypothalamus. Proc Natl Acad Sci USA 76:1448–1452

Lucke C, Hoffken B, von zur Mullen A (1976) Studies on the postponed growth hormone secretion following the infusion of somatostatin. Acta Endocrinol (Copenh) 82:460–466

Magoun H, Rhines R (1946) An inhibitory mechanism in the bulbar reticular formation. J Neurophysiol 9:165–171

Martin JB (1976) Brain regulation of growth hormone secretion. In: Martini L, Ganong WF (eds) Frontiers in neuroendocrinology, vol 4. Raven, New York, pp 129–168

Matsumoto J, Sogabe K, Hori-Santiago Y (1972) Sleep in parabiosis. Experientia 28: 1043–1044

McKelvy JF (1977) Biosynthesis of hypothalamic peptides. Adv Exp Med Biol 87: 77–98

McKelvy JF, Epelbaum J (1978) Biosynthesis, packing, transport, and release of brain peptides. In: Reichlin S, Baldessarini RJ, Martin JB (eds) The hypothalamus. Raven, New York, pp 195–211

McKelvy JB, Charli JL, Joseph-Bravo P, Sherman T, Loudes C (to be published) Cellular biochemistry of brain peptides biosynthesis, degradation, packaging, transport, and release. In: Motta M (ed) The endocrine function of the brain. Raven, New York

McNeill TH, Scott DE, Sladek JR Jr (1980) Simultaneous monoamine histofluorescence and neuropeptide immunocytochemistry: I. Localization of catecholamines and gonadotropin-releasing hormone in the rat median eminence. Peptides 1:59–68

Mendelson WB, Gillin JC, Pisner G, Wyatt RJ (1980) Arginine Vasotocin and sleep in the rat. Brain Res 182:246–249

Mendelson WB, Slater S, Gold P, Gillin JC (1980) The effect of growth hormone administration on human sleep: a dose response study. Biol. Psychiatry 15:613–618

Miselis RR, Shapiro RE, Hand PJ (1979) Subfornical organ efferents to neural systems for control of body water. Science 205:1022–1025

Monnier M, Hosli L (1964) Dialysis of sleep and waking factors in blood of the rabbit. Science 146:797–798

Monnier M, Hosli L (1965) Humoral transmission of sleep and wakefulness. 11. Hemodialysis of a sleep inducing humor during stimulation of the thalamic somnogenic area. Pfluegers Arch 282:60–75

Monnier M, Hatt AM (1971) Humoral transmission of sleep. V. New evidence from production of pure sleep hemodialyzate. Pfluegers Arch 329:231–234

Monnier M, Koller T, Graber S (1963) Humoral influences of induced sleep and arousal upon electrical brain activity of animals with crossed circulation. Exp Neurol 8:264–277

Monnier M, Hatt AM, Cueni LB, Schoenenberger GA (1972) Humoral transmission of sleep. VI. Purification and assessment of a hypnogenic fraction of "sleep dialyzate" (factor delta). Pfluegers Arch 331:257–265

Monnier M, Dudler L, Schoenenberger GA (1973) Humoral transmission of sleep. VIII. Effects of the "sleep factor delta" on cerebral, motor, and visceral activities. Pfluegers Arch 345:23–35

Monnier M, Dudler L, Gachter R, Maier PF, Tobler HJ, Schoenenberger GA (1977) The delta sleep inducing peptide (DSIP). Comparative properties of the original and synthetic nonapeptide. Experientia 33:548–552

Moore RY (1978) Central neural control of circadian rhythms. In: Ganong WF, Martini L (eds) Frontiers in neuroendocrinology, vol 5. Raven, New York, pp. 185–206

Moore RY (1979) The anatomy of central neural mechanisms regulating endocrine rhythms. In: Krieger DT (ed) Endocrine rhythms. Raven, New York, pp 63–87

Morin LP, Fitzgerald KM, Zucker I (1977) Estradiol shortens the period of hamster circadian rhythms. Science 196:305–307

Morrison A (1979) Brain stem regulation of behavior during sleep and wakefulnes. Prog Psychobiol 8:91–131

Moruzzi G (1972) The sleep-waking cycle. Ergb Physiol Biol Chem Exp Pharmakol 64:1–165

Moss RL (1979) Actions of hypothalamic hypophysiotropic hormones in the brain. Annu Rev Physiol 41:617–631

Moss RL, Riskind P, Dudley CA (1979) Effects of LHRH on sexual activities in animal and man. In: Collu R, Barbeau A, Ducharme JR, Rochefort VG (eds) Central nervous system effects of hypothalamic hormones and other peptides. Raven, New York, pp 345–366

Mullen PE, Jeffcoate WJ, Linsell C, Howard R, Rees LH (1979) The circadian variation of immunoreactive lipotropin and its relationship to ACTH and growth hormone in man. Clin Endocrinol (Oxford) 11:533–539

Nagasaki H, Iriki M, Indue S, Uchizono K (1974) The presence of a sleep promoting material in the brain of sleep-deprived rats. Proc Jap Acad 50:241–246

Nagasaki H, Iriki M, Uchizono K (1976) Inhibitory effect of the brain extract of sleep-deprived rats (BE-SDR) on the spontaneous discharges of crayfish abdominal ganglion. Brain Res 109:202–205

Neill JD (1980) LHRH and dopamine secretion into hypophyseal stalk blood. Effects of estrogen, mating, and nursing. In: Valverde-R C, Arechiga H (eds) Comparative aspects of neuroendocrine control of behavior. Karger, Basel Paris London New York, pp 192–217

Negoro H, Visessuwan S, Holland DR (1973) Unit activity in the paraventricular nucleus of female rats at different stages of the reproductive cycle and after ovariectomy, with or without oestrogen or progesterone treatment. J Endocrinol 59:545–558

Negro-Vilar A, Sanchez-Franco F, Kwiatkowski M, Samson WK (1979) Failure to detect radioimmunoassayable arginine vasotocin in mammalian pineals. Brain Res Bull 4:789–792

Nicoll CS (1980) Ontogeny and evolution of prolactin's functions. Fed Proc 39: 2563–2566

Nilaver G, Zimmerman EA, Wilkins J, Michaels J, Hoffman D, Silverman A-J (1980) Magnocellular hypothalamic projections to the lower brain stem and spinal cord of the rat. Neuroendocrinology 30:150–158

Ochs S (1977) Axoplasmic transport of peripheral nerve and hypothalamoneuro-hypophyseal systems. In: Porter JC (ed). Hypothalamic peptide hormones and pituitary regulation. Adv Exp Med Biol 87:13–40

O'Donohue TL, Miller RL, Pendleton RC, Jacobowitz DM (1979) A diurnal rhythm of immunoreactive -melanocyte-stimulating hormone in discrete regions of the rat brain. Neuroendocrinology 29:281–287

O'Malley BW, Roop DR, Lai EC, Nordstrom JL, Catteral JF, Swaneck GE, Colbert DA, Tsai M-J, Dugaiczyk A, Woo SLC (1979) The ovalbumin gene: organization, structure, transcription and regulation. Recent Progr Horm Res 35:1–46

Oswald I (1969) Human brain protein, drugs and dreams. Nature 223:893–897

Pacold ST, Hojvat S, Kirsteins L, Yarzagary L, Kisla J, Lawrence AM (1977) Brain growth hormone: evidence for the presence and production of biologically active GH-like immunoreactivity from the amygdaloid nucleus. Clin Res 25:299

Pappenheimer JR, Miller TB, Goodrich CA (1967) Sleep promoting effects of cerebrospinal fluid from sleep-deprived goats. Proc Nat Acad Sci 58:513–517

Pappenheimer JR, Koski G, Fencl V, Karnovsky ML, Krueger J (1975) Extraction of sleep promoting factor S from cerebrospinal fluid and from brains of sleep deprived animals. J Neurophysiol 38:1299–1311

Parker DC, Rossman LG, Vanderlaan EF (1974) Relation of sleep-entrained human release to REM and non REM cycles. J Clin Endocrinol Metab 38:646–651

Parker DC, Rossman LG, Kripke DF, Gibson W, Wilson K (1979) Rhythmicities in human growth hormone concentrations in plasma. In: Krieger DT (ed) Endocrine rhythms. Raven, New York, pp 143–173

Pavel S (1965) Evidence for the presence of lysive vasotocin in the pig pineal gland. Endocrinology 77:812–817

Pavel S (1979) Pineal vasotocin and sleep: involvement of serotonin containing neurons. Brain Res Bull 4:731–734

Pavel S, Matrescu L, Petrescu M (1973) Central corticotropin inhibition by arginine vasotocin. Neuroendocrinology 12:371–375

Pavel S, Goldstein E, Ghinea E, Calb M (1977a) Chromatographic evidence of vasotocin biosynthesis by cultures pineal ependymal cells from rat fetures. Endocrinology 100:205–208

Pavel S, Psatta, Goldstein R (1977b) Slow wave sleep induced in cats by extremely small amounts of synthetic and pineal vasotocin inyected into the third ventricle of the brain. Brain Res Bull 2:251–254

Pearse AGE (1976) Peptides in brain and intestine. Nature 262:92

Pearse AGE (1977) The diffuse neuroendocrine system and the APUD concept. Related "endocrine" peptides in brain, intestine, pituitary, placenta and anurain cutaneous glands. Med Biol 55:115

Pearse AGE, Takor-Takor T (1976) Neuroendocrine embryology and the APUD concept. Clin Endocrinol [Suppl] 5:2295

Pegram V, Hammond D, Bridgers W (1973) The effects of protein synthesis inhibition on sleep in mice. Behav Biol 9:377–382

Pelletier G, Desy L (1979) Localization of ACTH in the human hipothalamus. Cell Tissue Res 196:525–530
Pelletier G, Leclerc R (1979) Immunohistochemical localization of adrenocorticotropin in the rat brain. Endocrinology 104:1426–1433
Perez-E B, Valverde-R C (to be published) The diffuse neuroendocrine system and its pathology: antecedents, present concepts, perspectives. Metabolism
Peters JA, Santa-Cruz F, Rubin RT (1980) Neuroendocrine responses to TRH during REM and slow-wave sleep in man (Abstr 634). The Endocrine Society 62nd Annual Meeting. Endocrine Soc, Washington, p 233
Petitjean F, Buda C, Janin M, David M, Jouvet M (1979) Effets du Chloramphenicol sur le sommeil du chat – Comparison avec le thiamphénicol, l'erythromycine et l'erytrhomycine et l'oxytétracycline. Psychopharmacologie 66:147–153
Phillips MI, Weyhenmeyer J, Felix D, Ganten D, Hoffman WE (1979) Evidence for an endogenous brain renin-angiotensin system. Fec Proc 38:226–266
Pieron H (1913) Le probleme physiologique du sommeil. Masson, Paris
Polak JM (1979) The diffuse neuroendocrine system. Studies of this newly discovered controlling system in health and disease. J Histochem Cytochem 27:1398–1400
Polc P, Schneeberger J, Haefely W (1978) Effect of the delta sleep inducing peptide (DSIP) on the sleep-wakefulness cycle of cats. Neurosci Lett 9:33–36
Polet RJ, Levey GS (1980) Principles of membrane receptor physiology and their application to clinical medicine. Ann Int Med 92:663–680
Purpura D (1956) A neurohumoral mechanism of reticular cortical action. Am J Physiol 186:50–54
Quabbe HJ (1977) Chronobiology of growth hormone secretion. Chronobiologia 4: 217–246
Ramirez G (1973) Synaptic plasma membrane protein synthesis: selective inhibition by chloramphenicol *in vivo.* Biochem Biophys Res Commun 50:452–458
Ramm P (1979) The locus coeruleus, catecholamines and REM sleep: a critical review. Behav Neural Biol 25:415–448
Reichlin S (1980) Peptides in neuroendocrine regulation. In: Bloom F (ed) Peptides: integrators of cell and tissue function. Raven, New York, pp 235–250
Reichlin S, Connolly J, Greens GL, Jensen EV, Robbins RJ (1980) Histochemical localization of estrophilin-like immunoreactivity in rat hypothalamic cell cultures using anti-calf uterine estrophilin antibody (Abstr 121). The Endocrine Society 62nd Annual Meeting. Endocrine Soc, Washington, p 105
Renaud LP, Pittman QJ, Blume HW (1978) Neurophysiology of hypothalamic peptidergic neurons. In: Fuxe K, Hökfelt T, Luft R (eds) Central regulation of the endocrine system. Plenum, New York London, pp 119–136
Renaud LP, Blume HW, Pittman QJ, Lamour Y, Tan AT (1979 a) Thyrotropin releasing hormone selectively depresses glutamate excitation of cerebral cortical neurons. Science 205:1275–1277
Renaud LP, Pittman Q, Blume H, Lamour Y, Arnaud E (1979b) Effects of peptides on central neuronal excitability. In: Collu R, Barbeau A, Ducharme JR, Rochefor JC (eds) Central nervous system effects of hypothalamic hormones and other peptides. Raven, New York, pp 147–162
Rezek M, Havlicek V, Hughes K, Friesen H (1976) Cortical administration of somatostatin (SRIF): effect on sleep and motor behavior. Pharmacol Biochem Behav 5:73–77
Rillema JA (1980) Mechanism of prolactin action. Fed Proc 39:2593–2598
Ringle DA, Herndon BL (1968) Plasma dialyzates from sleep-deprived rabbits and their effect on the electrocorticogram of rats. Pfluegers Arch 303:344–349
Ringle D, Herndon B (1969) Effects on rats of CSF from sleep deprived rabbits. Pfluegers Arch 30:320–328
Roger LJ (1979) Evidence for T4-GH interaction during brain development. Nature 282:414–415

Rojas-Ramirez JA, Aguilar-Jimenez E, Posadas-Andrews A, Bernal-Pedraza J, Drucker-Colín RR (1977) The effects of various protein synthesis inhibitors on the sleep-wake cycle of rats. Psychopharmacol 53:147–150

Rosenbloom AA, Fisher DA (1974) Radioimmunoassay of arginine vasotocin. Endocrinology 95:1726–1732

Rossier J, Pittman Q, Bloom F, Guillemin R (1980) Distribution of opiod peptides in the pituitary: a new hypothalamic-pars nervosa enkephalinergic pathway. Fed Proc 39:2555–2560

Rubin RT, Poland RE, Ravessound F, Gouin PR, Tower BB (1975) Antidiuretic hormone: episodic nocturnal secretion in adult men. Endocr Res Commun 2: 461–469

Rusak B, Zucker I (1979) Neural regulation of circadian rhythms. Physiol Rev 59: 449–526

Sachs H, Pearson D, Nureddin H (1975) Guinea pig neurophysin: isolation, developmental aspects, biosynthesis in organ culture. Ann NY Acad Sci 248:36–45

Sakai K, Sastre JP, Salvert D, Touret M, Tohyama M, Jouvet M (1979) Tegmento reticular projections with special reference to the muscular otonia during paradoxical sleep: an HRP study. Brain Res 176:233–254

Sassin J, Parker DC, Johnson LC, Rossman LG, Male JW, Gotlin RW (1969) Effects of slow wave sleep deprivation in human growth hormone – release in sleep: preliminary study. Life Sci 8:1299–1307

Sastre JP (1978) Effects des lesions du tegmentum pontique sus l'organization des etats de sommeil dies la chat. PhD dissertation, Université Claude Bernard, Lyon

Sastre JP, Jouvet M (1979) Le comportement oniseque du chiat. Physiol Behav 22: 979:989

Scharrer B (1978) Peptidergic neurons. Facts and trends. Gen Comp Endocrinol 34:50–62

Scherrer H, Seidah NG, Benjannet S, Crine P, Cretien M (1979) Biosynthese de beta-lipotropin dans le tissue cerebral in citro. Crit Rev Acad Sci (Paris) 288:543–546

Schnedorf JG, Ivy AC (1939) An examination of the hypnotoxin theory of sleep. Am J Physiol 125:191–205

Schnitzer SB, Ross S (1960) Effects of physiological saline injection on locomotor activity in C57 BL/6 mice. Psychol Rep 6:351–354

Schoenenberger GA, Monnier M (1977) Characterization of a delta-electroencephalogram (sleep) inducing peptide. Proc Nat Acad Sci USA 74:1282:1286

Schoenenberger GA, Cueni LB, Monnier M, Hatt AM (1972) Humoral transmission of sleep. VII. Isolation and physico-chemical characterization of the "sleep inducing factor delta." Pfluegers Arch 338:1–17

Schoenenberger GA, Maier PF, Tobler JH, Monnier M (1977) A naturally occurring delta-EEG enhancing nonapeptide in rabbits. X. Final isolation characterization and activity test. Pfluegers Arch 369:99–109

Schoenenberger GA, Manier PF, Tobler HJ, Wilson K, Monnier M (1978) The delta EEG (sleep) inducing peptide (DSIP). XI. Amino acid analysis, sequence, synthesis and activity of the nonapeptide. Pfluegers Arch 378:119–129

Schultz WJ, Bronfield MS, Kizlowski GP (1977) The hypothalamo-choroidal tract. 11. Ultrastructural response of the coroid plexus to vasopressin. Cell Tissue Res 178:129–141

Schwarz WJ, Gainer H (1977) Suprachiasmatic nucleus: use of 14 D-labeled deoxyglucose uptake as a functional marker. Science 197:1089–1091

Seif SM, Robinson AG (1979) Rhythms of the posterior pituitary. In: Krieger DT (ed) Endocrine rhythms. Raven, New York, pp 187–201

Smith OA, Astley CL, DeVito JL, Stein JM, Walsh KE (1980) Functional analisis of hypothalamic control of the cardiovascular responses accompanying emotional behavior. Fed Proc 39:2487–2494

Solis H, Fernandez-Guardiola A, Valverde-R C (1979) Neuropharmacologic and neuroendocrine interrelations of human sleep. In: Drucker-Colín R, Shkurovich M, Sterman MB (eds) The functions of sleep. Academic Press, New York, pp 147–170

Spanis CW, Gutierrez MC, Drucker-Colin PR (1976) Neurohumoral correlates of sleep: further biochemical and physiological characterization of sleep perfusates. Pharmacol Biochem Behav 5:165–173

Stemberger LA, Hoffman GE (1978) Immunocytology of luteinizing hormone releasing hormone. Neuroendocrinology 25:111–128

Steriade M, Hobson JA (1976) Neural activity during the sleep waking cycle. Prog Neurobiol 6:155–376

Stern WC, Morgane PJ (1977) Sleep and memory: effects of growth hormone on sleep, brain biochemistry and behavior. In: Drucker-Colín RR, McGaugh JL (eds) Neurobiology of sleep and memory. Academic Press, New York, pp 373–410

Stern WC, Miller FP, Cox RH, Maickerl RP (1971) Brain norepinephrine and serotonin levels following REM sleep deprivation in the rat. Psychopharmacoly 22:50–55

Stern WC, Morgane PJ, Panksepp J, Solovick AJ, Jalowiec JE (1972) Elevation of REM sleep following inhibition of protein synthesis. Brain Res 47:254–258

Stern WC, Jalowiec E, Shabshalowtiz H, Morgane PJ (1975) Effects of growth hormone on sleep-waking patterns in cats. Horm Behav 6:189–196

Stumpf WE (1970) Estrogen-neurons and estrogen-neuron systems in the periventricular brain. Am J Anat 129:207–218

Takahashi Y (1979) Growth hormone secretion related to the sleep and waking rhythm. In: Drucker-Colín R, Shkurovich M, Sterman MB (eds) The functions of sleep. Academic Press, New York, pp 113–145

Takahashi Y, Kipnis DM, Daughaday WH (1968) Growth hormone secretion during sleep. J Clin Invest 47:2079–2090

Takahashi Y, Ebihaba S, Nakamura Y, Takahashi K (1978) Sleep related growth hormone secretion in dogs after 8 h forced wakefulness. In: Tsukahara N, Kubata K, Yagi K (eds) Integrative control functions of the brain. Elsevier, Amsterdam, pp 389–391

Tata JR (1968) Hormonal regulation of growth and protein synthesis. Nature 219: 331–337

Terry LC, Martin JB (1978) Hypothalamic hormones: subcellular distribution and mechanisms of release. Annu Rev Pharmacol Toxicol 18:111–123

Tobler I, Borbely A (1980) Effect of delta sleep inducing peptide (DSIP) and arginine vasotocyn (AVT) on sleep and motor activity in the rat. Waking Sleeping 4:139–153

Toubeau G, Desclin J, Parmentier M, Pasteels JL (1979a) Compared localizations of prolactin-like material and adrenocorticotropin immunoreactivities within the brain. Neuroendocrinology 29:374–384

Toubeau G, Desclin J, Parmentier M, Pasteels JL (1979b) Cellular localization of a prolactin-like antigen in the rat brain. J Endocrinol 83:261–266

Vale W, Rivier C, Brown M (1977) Regulatory peptides of the hypothalamus. Annu Rev Physiol 39:473–528

Valverde-R C (to be published) La integración neuroendocrina y el binomio vigilia-sueño como un sistema de intercomunicación. Rev Asoc Psiq Méx

Valverde-R C, Jurado J, Ruiz-Primo E, Ruiz-Juvera A, Maisterrena JA, Fernández-Guardiola A (1980) Hormones and sleep rhythms in endocrine dysfunction. Front Horm Res, pp 156–175

Vanderleghen JJ, DeMey J, Lostra F, Giles C (1979) Localization of gastrin-cholecystokinin-like peptides in the brain and hypophysis of the rat. Acta Neurol Belg 79:62–63

Vandesande F, Dierickx K, DeMey J (1975) Identification of the vasopressin-neurophysin producing neurons of the rat suprachiasmatic nuclei. Cell Tissue Res 156: 377–380

Van Houten M, Posner BI, Kopriwa B, Brawer JR (1980) Insulin binding sites localized to nerve termin als in rat median eminence and arcuate-nucleus. Science 207: 1081–1083

Van Vugt DA, Meites J (1980) Influence of endogenous opiates on anterior pituitary function. Fed Proc 39:2533–2538

Vijayan E, Samson WK, Said SI, McCann SM (1979) Vasoactive intestinal peptide: evidence for a hypothalamic site of action to release growth hormone, luteinizing hormone and prolactin in conscious ovariectomized rats. Endocrinology 104: 53–57

Viveros OH, Diliberto EJ, Hazum E, Chang K-J (to be published) Enkephalins as possible adrenomedullary hormones: storage, secretion and regulation of synthesis. Adv Biochem Psychopharmacol

Wagner DR, Weitzmann ED (1980) Neuroendocrine secretion and biological rhythms in man. Psychiatr Clin North Am 3:223–250

Watson SJ, Richard CW III, Ciaranello RD, Barchas JD (1980) Interactions of opiate peptide and noradrenalin systems: light microscopic studies. Peptides 1:23–30

Weindl A, Sofroniew MV (1976) Demonstration of extrahypothalamic peptide secreting neurons. A morphologic contribution to the investigation of psychotropic effects of neurohormones. Pharmakopsychiatr Neuropsychopharmakol 9:226–234

Weiner RI, Ganong WF (1978) Role of brain monoamines and histamine in regulation of anterior pituitary secretion. Physiol Rev 58:905–976

Weitzman ED (1980) Biological rhythms and hormonal secretion patterns. In: Krieger DT, Hughes JC (eds) Neuroendocrinology. Sunderland, Mass., Sinauer Associates, pp 85–92

Weitzman ED, Czeisler CA, Moore-Ede MC (1979) Sleep-wake neuroendocrine and body temperature circadian rhythms under entrained and non-entrained (free running) conditions in man. In: Suda M, Hayaishi O, Nakagawa H (eds) Biological rhythms and their central mechanism. Elsevier/North-Holland Biomedical Press, Amsterdam Oxford New York, pp 199–227

Zieglgansberger W (1980) Peptides in the regulation of neural function. In: Bloom F (ed) Peptides: integrators of cell and tissue function. Raven, New York, pp 219–234

Zimmerman EA (1976) Localization of hypothalamic hormones by immunocytochemical techniques. In: Martin L, Ganong WF (eds) Frontier in neuroendocrinology. Raven, New York, pp 25–62

Zimmerman EA, Antunes JL (1976) Organization of the hypothalamo-pituitary system: current concepts from immunohistochemical studies. J Histochem Cytochem 24:807–815

Zucker I, Cramer CP, Bittman EL (1980) Regulation by the pituitary gland of circadian rhythms in the momster. J Endocrinol 85:17–25

Zwick M (1977) Some analogies of hierarchical order in biology and linguistics. In: Klir GJ (ed) Applied general systems research. Plenum, New York London, pp 521–530

[illegible]

Sleep Regulation: Circadian Rhythm and Homeostasis

ALEXANDER A. BORBÉLY[1]

Contents

Abbreviations

DD: continous darkness; 5-HT: 5-hydroxy tryptamine; τ: period of circadian rhythm; LD: light-dark; NREMS: non-REM sleep; REM: rapid eye movement; REMS: REM sleep; SCN: suprachiasmatic nucleus; SD: sleep deprivation; SWS: slow wave sleep; PCPA: p-chlorophenylalanine

1 Introduction

W. R. Hess characterized sleep as "... the expression of a predominance of the trophotropic component of the autonomous nervous system and a preventive measure against exhaustion ..." (Hess 1965). His concept of alternating trophotropic and ergotropic states resembles the present-day view of a circadian rest-activity rhythm. The trophotropic state and the circadian rest state have in common the predominance of physiological processes subserving energy conservation and restoration. They include, in addition to sleep, lipolytic processes of energy metabolism (Le Magnen et al. 1968; Danguir and Nicolaidis 1980) coupled to a low rate of feeding and drinking (e.g. Borbély 1977) and a low level of body temperature (Eastman 1980). The recognition

1 Institute of Pharmacology, University of Zurich, Zurich, Switzerland

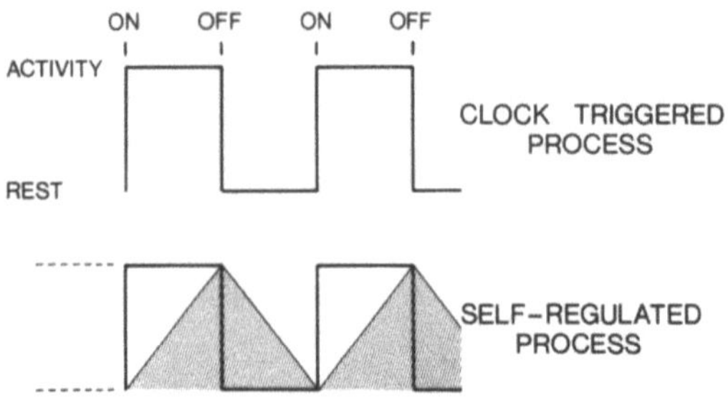

Fig. 1. Schematic representation of the circadian rest-activity cycle as a clock-triggered process and as a self-regulated process. Activity is indicated by the *upper level*, rest by the *lower level* of the square wave. The *interrupted horizontal lines* in the *lower diagram* represent thresholds for the regulatory variable whose level is indicated by the upper limit of the *hatched area*

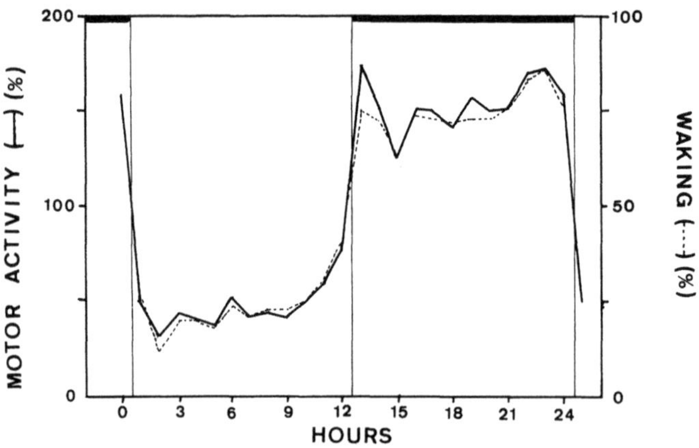

Fig. 2. Daily pattern of motor activity *(solid curve)* and waking *(interrupted curve)* in the rat. The *curves* interpolate the hourly mean values (19 days, ten rats). Motor activity (scale on *left ordinate*) is expressed relative to the mean daily activity level of individual animals (=100%), waking (scale on *right ordinate*) is expressed as a percentage of recording time. Dark periods are indicated by *bars* below *top abscissa* and delimited by *vertical lines* (from Borbély and Neuhaus 1978a)

of the integrative functions of the autonomic nervous system was at the root of the ergotropic-trophotropic state concept. The discovery of the existence of a central circadian oscillator was the major event in circadian rhythm research. Thus it became clear that the various physiological processes occurring typically during the circadian rest-phase are not merely a consequence of behavioural rest or sleep, but are under the direct control of a circadian pacemaker. The evidence was obtained from animals and people who lived under schedules without 24-h time-cues, and showed dissociations of their sleep-wake rhythm from the rhythms of body temperature or corticosteroid secretion, although sleep-related components were still evident (Weitzman et al. 1979; Eastman 1980). The results of rhythm research have therefore not only given a new significance to the concept of alternating ergotropic and trophotropic states, but have also shed light on the intrinsic control mechanisms of sleep and waking.

The upper part of Fig. 1 represents a simplified diagram of the circadian rest-activity cycle and its control by a circadian pacemaker. The oscillator is shown to trigger at regular intervals the onset and end of the activity period. In the lower diagram, the same rest-activity cycle is generated by a self-regulated process. The upper limit of the shaded area corresponds to the level of a regulatory variable which increases as a function of activity time. When it reaches an upper threshold, activity is switched off. The progressive decline of the regulatory variable during the rest period leads to the rest-activity transition as a lower threshold is reached. This self-regulated process may be regarded as a homeostatic mechanism which serves to maintain an essential, sleep-related variable within physiological limits. Homeostasis was defined by Cannon (1939) as "The coordinated physiological processes which maintain most of the steady states in the organism ...." Cannon attributed an essential role to the autonomic nervous system in homeostatic regulation. In the present review, the concept of homeostasis is extended to the regulation of sleep. Sleep and its substates are apparently regulated relative to an internal reference level, since sleep deprivation leads to a compensatory increase, whereas excess sleep lowers sleep propensity. However, in contrast to most other physiological processes that are subject to homeostatic regulation, the function of sleep is unknown. The analysis of the regulatory mechanisms is therefore also undertaken in the hope that insight may be gained into the functional significance of sleep.

2 The Sleep-Wake Cycle as a Circadian Rhythm

When man or animals live in an environment without time-cues, their sleep-wake cycle continues to exhibit a well-defined rhythm whose period (τ) is usually somewhat different from 24 h (e.g. rat: Borbély and Neuhaus 1978b; man: Wever 1979). The sleep-wake cycle constitutes therefore a true circadian rhythm which is controlled by an intrinsic oscillator. Research in circadian activity rhythms is based heavily on motor activity recordings, whereas long-term measurements of sleep have been rarely undertaken. Waking is a prerequisite for motor activity, although motor activity is not always present during waking. Particularly when wheel-running is used as the activity measure, inactivity cannot be equalled to rest or sleep. However, a close correspondence between the daily pattern of waking and motor activity is obtained, if movement activity is measured by transducers under the cage (Fig. 2). Long-term recordings of the rest-activity rhythm obtained by this method may provide therefore a close approximation of the sleep-wake rhythm.

Circadian activity rhythms recorded under conditions of continuous darkness (DD) are characterized by a high stability of τ. This is apparent from the records shown in Figs. 3 and 4 where the onset and offset times of the activity periods show a close correspondence with the average period τ. The stability of the circadian activity rhythm of rodents has been documented in detail (Pittendrigh and Daan 1976).

An outstanding feature of the circadian rest-activity rhythm is its insensitivity to procedures that are known to have profound effects on sleep. Figure 3 illustrates an experiment in which a rat maintained in DD was sleep deprived for 24 h. The marked changes of sleep during the recovery period are illustrated and discussed in Sect. 3, and have been presented in detail elsewhere (Borbély and Neuhaus, 1979). The records in Fig. 3 show that the sleep-deprivation (SD) procedure neither affected τ nor induced

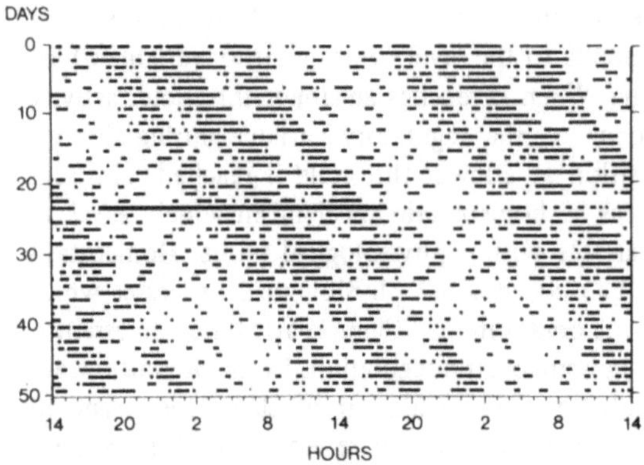

Fig. 3. Effect of sleep-deprivation on the circadian rest-activity rhythm of a rat maintained in continuous darkness. Suprathreshold activity values (recorded by a mechano-electric transducer under the cage for consecutive 15-min periods) are indicated by *horizontal lines.* Each day has been plotted twice to facilitate inspection of the circadian rhythm. The *continuous line* on day 24 indicates the 24-h sleep-deprivation period

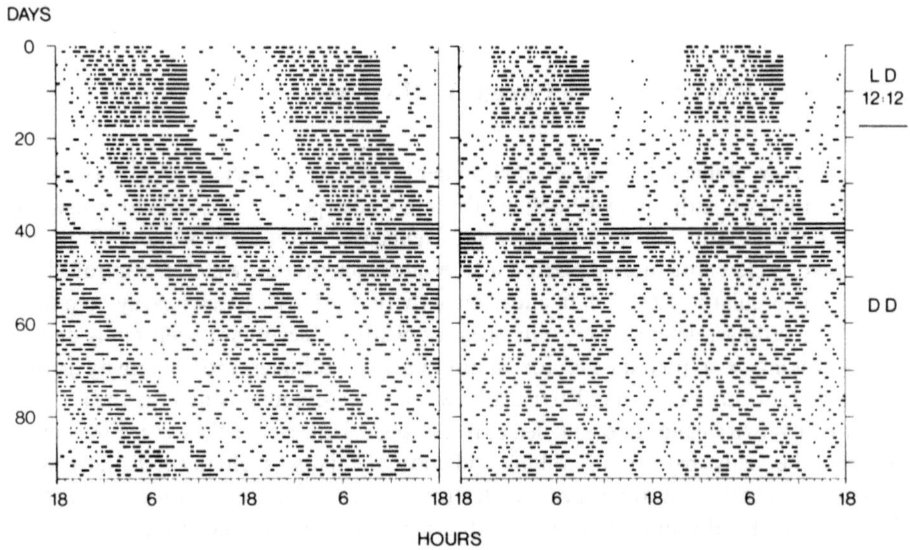

Fig. 4. Effect of social interaction on the circadian rest-activity rhythm. Activity records of two rats maintained in separate cages under entrained conditions *(LD 12:12)* and in continuous darkness *(DD)* up to day 40. Day 40–50: Social interaction period in which both animals were maintained in the same cage. Day 51–88: Recording in separate cages. Activity is plotted as in Fig. 3. Note the lack of effect of social interaction on period (τ) and phase of the circadian activity rhythm

a phase shift in the circadian activity rhythm. Analogous results have been obtained in eight other rats (Borbély and Tobler, unpublished work).

Figure 4 provides an illustration for the striking insensitivity of the rest-activity rhythm to social influences. The animals were first kept in individual cages under entrained conditions (LD 12:12). After "release" into DD they exhibited a typical free-running circadian activity rhythm with individually different τ. To test for interaction between the rhythms during social contact, the two animals were placed into the same cage for a 10-day period (day 40–50). Then they were separated and recorded again in individual cages during the subsequent weeks. It is evident from Fig. 4 that the social contact period neither affected τ nor induced a phase shift in the circadian activity rhythm of the two animals. This experiment was performed in 14 rats and revealed no significant effect of social interaction on the parameters of the rest-activity rhythm (Borbély et al., in preparation).

The experiments illustrate the stability of the circadian activity rhythm during schedules which are likely to influence sleep. However, there are also procedures which do affect the circadian rhythm while leaving the sleep state distribution largely unchanged. Thus, in accordance with Aschoff's rule, τ of the circadian sleep-wake rhythm of rats was significantly longer in continuous light than in continuous darkness (Borbély and Neuhaus 1978b), while the basic circadian pattern and the percentage of sleep states were similar under the two conditions. A more drastic dissociation between the sleep-wake rhythm and the sleep state distribution was seen in animals with lesions of the suprachiasmatic nucleus (SCN). In this preparation, the circadian sleep-wake rhythm was no longer detectable, whereas the daily percentage of total sleep, non-REM sleep (NREMS) and REM sleep (REMS) remained at the prelesion control level (Ibuka and Kawamura 1975; Ibuka et al. 1977). Experiments are in progress to investigate whether homeostatic sleep regulation is still present in SCN-lesioned animals. The experiments described in this section indicate that the circadian sleep-wake rhythm and homeostatic sleep regulation may be separately manipulated, and may be based therefore on independent mechanisms.

3 Sleep as a Homeostatic Process

3.1 Studies in Man

After a prolonged period of sleep deprivation (SD) the duration of recovery sleep is in general longer than the usual sleep period. However, the excess sleep time is modest in comparison to sleep loss (e.g. Gulevich et al. 1966). This raises the question whether a specific sleep state may preferentially compensate the sleep deficit. According to such an assumption, a compensatory prolongation of sleep time could be avoided by an increase in "sleep intensity." There is evidence that slow wave sleep (SWS; stages 3 and 4 of NREMS) represents a "high-intensity" fraction of NREMS and that it may reflect a restorative sleep process. One indication is based on the predominance of SWS in the first part of the sleep period. This is illustrated by data from a human study (Fig. 5: top diagram) where stage 4 is limited to the first NREM-REM sleep cycle, while stage 3 is restricted to the first two cycles. The second part of the sleep period is dominated by stage 2 and REMS. The four stages of NREMS are based on specified, yet nevertheless arbitrary EEG criteria (Rechtschaffen and Kales 1968) which are applied to subdivide the essentially continuous NREMS process into discrete substrates. Power density plots of the various EEG frequency bands reflect more adequately

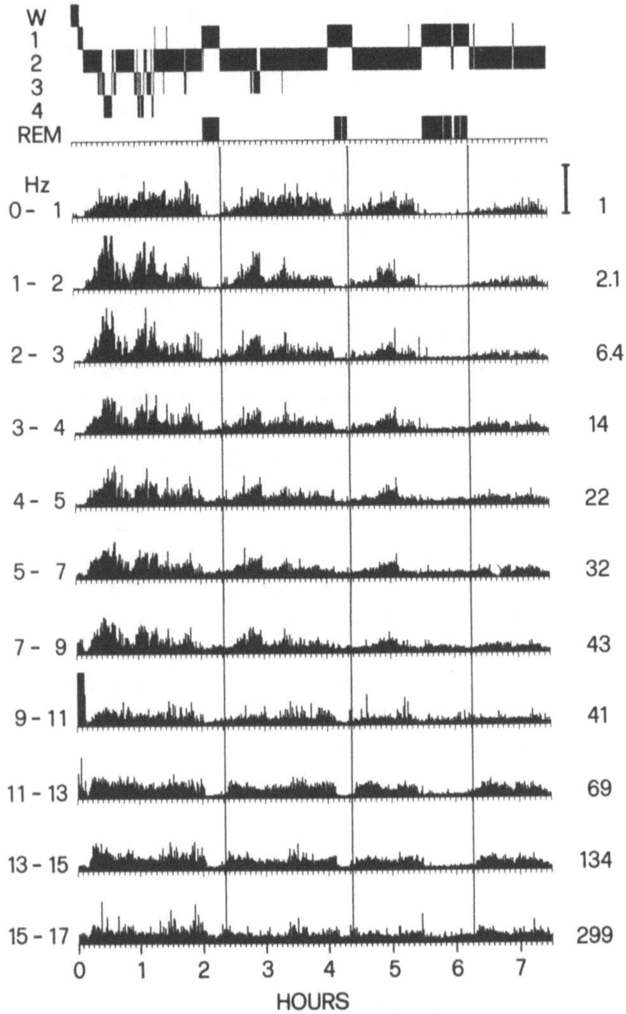

Fig. 5. Sleep states and EEG power density plots of a human baseline night. Sleep states *(top)* are plotted for 30-s epochs (*W*, waking: 1–4: non-REM sleep stages; *REM:* REM sleep) (note that REM sleep epochs are also plotted as stage 1 sleep). Power density of the various frequency bands is plotted for 1-min segments. The lower limits of the frequency bands are 0.25 Hz higher than indicated on the left (e.g. 0–1 Hz corresponds to 0.25–1.00 Hz). The calibration mark at the right of the top frequency band corresponds to 200 $(\mu V)^2/0.25$ Hz for scale factor 1. The other scale factors indicated on the *right* represent relative values which were chosen to make the area of the plots equal. The non-REM-REM sleep cycles are delimited by *vertical lines* (unpublished data from Borbély, Baumann, Brandeis, Strauch, and Lehmann 1981 a)

the continuous aspect of the sleep process and allow therefore a more detailed analysis of its changes. The plots shown in Fig. 5 are based on power spectra computed with a Fast Fourier Transform routine for consecutive 4-s epochs which were then averaged over 1-min segments (see Borbély et al. 1981 a for details). The four NREM-REM sleep

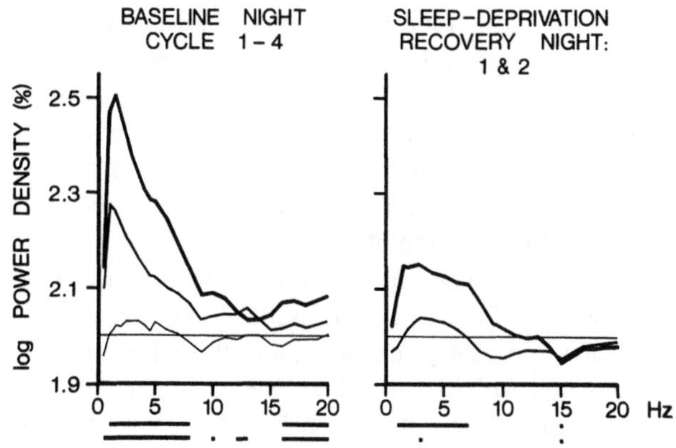

Fig. 6. Frequency distribution of relative EEG power density for baseline nights *(left)* and the two nights following sleep deprivation *(right)*. The *curves* connect the mean values from eight subjects computed for 0.5 Hz bins between 0.25 and 5.0 Hz, and for 1 Hz bins between 5.25 and 25.0 Hz. In the *left panel,* values for non-REM-REM sleep cycles 1,2, and 3 are indicated by *curves of decreasing thickness.* For each subject the value for sleep cycle 4 has been defined as 100% (indicated by *horizontal line*) for each frequency band. Values of the preceding cycles are expressed relative to the corresponding value of cycle 4. In the *right panel,* the values of the baseline nights were defined as 100%. The curves representing recovery night 1 *(thick line)* and 2 *(thin line)* are expressed relative to the corresponding baseline values. Lines below *abscissa* indicate significant differences between frequency bands ($P<0.05$; 2-sided paired Wilcoxon test). *Left panel: upper lines* indicate difference between cycle 1 and 2, *lower lines* between cycle 2 and 3. *Right panel: upper lines* indicate difference between baseline and recovery night 1; *lower lines* between baseline and recovery night 2 (modified from Borbély et al. 1981 a)

cycles (Feinberg and Floyd 1979) are delimited by vertical lines. The large peaks in the low frequency bands correspond to SWS. With the exception of the lowest frequency band, the delta peak in the first cycle typically exceeds the peak in the second cycle (Church et al. 1975). As the plots illustrate, the SWS peaks are not limited to the delta band, but extend up to 9 Hz. An alpha peak (9–11 Hz) is seen during waking prior to sleep onset. The high values in the 11–15 Hz band during stage 2 represent spindle activity. The progressive change in power density between the successive sleep cycles is evident also from the curves shown in Fig. 6 (left part) which represent standardized average values from eight subjects (16 baseline nights). The significant decrease from cycle 1 to cycle 3 encompasses the low frequency range up to 8 Hz as well as the high frequency range (16–20 Hz).

Is the predominance of SWS in the first cycle a consequence of the preceding waking period? In this case, sleep deprivation should lead to an increase of SWS in the recovery night, an effect that should be particularly prominent in the first NREM-REM sleep cycle. This prediction was confirmed in a recent study in which subjects were kept awake for 40.5 h (Borbély et al. 1981 a). In comparison to the baseline night, power density in the 1–7 Hz range was significantly increased in the first recovery night, and a residual effect was still evident in the subsequent night (Fig. 6, right part). The effect was most prominent in the first cycle. Furthermore, the increase in power density in the low frequency bands was observed not only for the total sleep period (as shown in Fig. 6), but was present also in SWS, stage 4, stage 2 and REMS when these stages were analysed separately. The study also demonstrates therefore the occurrence of significant changes *within* separate sleep stages. The high initial power density in the low frequency range, and its progressive decline seem to reflect a general sleep process which is not restricted to specific sleep states, and whose initial level is determined by the length of prior waking.

Effects of sleep deprivation on sleep stages and EEG have been reported previously. Thus Blake and Gerard (1937) described more than 4 decades ago the early occurrence of the slow wave sleep maximum and the subsequent decline across the sleep period. They also recognized the enhancement of slow waves after sleep deprivation. Later reports of the close relationship between SWS and prior waking include the papers of Berger and Oswald (1962), Williams et al. (1964), Webb and Agnew (1971), and Moses et al. (1975). Our own study agrees with that of Nakazawa et al. (1978) that sleep loss of one night enhances SWS, but leaves the percentage of REMS unchanged. The high priority of SWS emerges also from experiments in which sleep time was curtailed over prolonged time periods (Webb and Agnew 1974). Under these conditions, SWS was maintained at the control level or even increased, whereas REMS fell far below the baseline level.

Information about sleep-regulating mechanisms may be gained not only from studies in which sleep is completely or partly prevented, but also from experiments in which subjects are deprived of a particular substate. Soon after the discovery of REMS as a separate sleep state it was shown that its selective deprivation gives rise to a selective compensatory response during recovery sleep (Dement 1960). The analogous selective compensation of stage 4 was demonstrated later (Agnew et al. 1964). The results of these experiments suggest the existence of separate regulatory mechanisms for NREMS and REMS. However, since on the basic of the previous arguments, stage 4 sleep may not represent a state of its own, but one extreme along the NREMS continuum (see Figs. 5 and 6), a stage 4 deficit should be partly compensated by other NREMS stages. This assumption of a homeostatic regulation extending across the entire range of NREMS is supported by results from two experiments. In the first study, a 2-day total SD period followed immediately by 2 days of selective stage 4 deprivation induced less stage 4 rebound than the 2-day total SD alone (Moses et al. 1975). The data suggest that the selective stage 4 deficit was partly compensated for by other NREMS stages. In a second sudy, Feinberg et al. (1979) suggested that the reduction of stage 4 sleep by the hypnotic drug flurazepam may be spurious since the delta waves of this sleep stage were redistributed to stage 2 sleep. The continuous change of EEG parameters within NREMS in our own study (Figs. 5, 6) also supports the notion that the stages of NREMS may reflect different levels of intensity of the same functional process.

3.2 Studies in Animals

If sleep and its regulatory mechanisms are basically invariant across mammalian species, the features summarized in the preceding section for man should be valid also for animals. With rare exceptions, the two sleep states NREMS and REMS have been

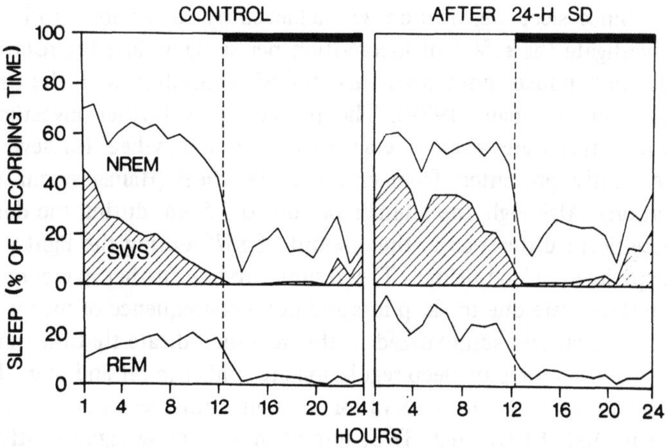

Fig. 7. Effect of sleep deprivation *(SD)* on sleep state pattern in the rat. The curves represent mean hourly values (n=6) expressed as percent of recording time. Slow wave sleep *(SWS)* indicated by the *hatched area* was defined on the basis of the zero-crossing analysis as the non-REM sleep fraction with the lowest 30% of zero-crossing values on baseline days. Dark period is delimited by a *bar* below the top *abscissa* and an *interrupted vertical line* (modified from Borbély and Neuhaus 1979)

identified in all mammals so far examined. The rat will be used to illustrate various aspects of sleep regulation in animals.

In many animals sleep occurs preferentially within the light or the dark part of the day-night cycle. In the rat, a nocturnal animal, sleep is present during approximately 80% of the light hours and during 20% of the dark hours (Fig. 2). REMS occupies 15%–20% of total sleep time, and exhibits a slight increasing trend within the 12-h light period (Fig. 7, left part). A selective REMS deprivation is followed by a selective REMS rebound (e.g. Morden et al. 1967). While basic aspects of REMS regulation are comparable in rat and man, the situation for NREMS appears at first sight to be different. Thus the slight decreasing trend of NREMS within the daily sleep (light) phase of the rat (Fig. 7) does not indicate that this sleep state is markedly influenced by the preceding dark-time waking period. However, a further analysis of NREMS according to EEG frequency criteria revealed a marked decreasing trend of the SWS fraction (Fig. 7; hatched area) which is quite comparable to the situation in man. The rat data were originally based on zero-crossing analysis, and were later confirmed by spectral analysis which showed also a decreasing trend in the delta band (1.25–4.50 Hz; Borbély 1980 a; Borbély et al. 1981 b).

To confirm that the high level of SWS in the first part of the light period is causally related to the preceding dark-time waking period, rats were subjected to SD (Borbély and Neuhaus 1979). SD over 12 h or 24 h was carried out by placing the animal into a slowly rotating cylinder (one rotation per 45 s) to which it had been adapted and where it had free access to food and water. Vigilance during SD could be monitored by EEG telemetry. Figure 7 (right part) shows that SWS was massively enhanced after 24-h SD, whereas light-time NREMS fell even below the baseline level. A marked REMS rebound was seen.

Since sleep deprivation was achieved by forced locomotion, it was mandatory to investigate the effect of locomotion per se. Doubling the rotation rate of the cylinder did not induce more SWS than the SD procedure with the usual rotation rate (Borbély and Neuhaus 1979). The problem was further investigated in an experiment where rats were given free access to a running wheel for several days, and were subsequently prevented from entering the wheel (Hanagasioglu and Borbély, in preparation). Although the animals ran up to 7.5 km during the dark hours, there was no significant difference in the amount of SWS during the light period between the two experimental conditions. The results provide further evidence that the effects of SD on sleep were due to sleep loss and not a consequence of motor activity.

The studies summarized in this section indicate that the daily sleep pattern as well as main features of sleep regulation are similar in rat and man. The similarity is further underscored by the observation that in both species short periods of SD (rat: 12 h; man: 40.5 h) enhance SWS, but do not increase significantly the REMS level (Borbély and Neuhaus 1979; Borbély et al. 1981a). Moreover, the analysis of daytime naps in man (Karacan et al. 1970) and of dark time sleep in rat (Borbély and Neuhaus 1979) revealed for both species an increasing propensity of SWS with the progression of the waking phase. Our own observations in the rat are in agreement with the results of the elegant studies by Rosenberg and coworkers (1976) and Friedman and coworkers (1979). Moreover, experiments performed in other species also confirm the main aspects of the data obtained for rat and man. Thus a predominance of SWS in the early part of the daily sleep phase was observed in the monkey (Reite et al. 1965; Crowley et al. 1972). The increase of SWS as a function of prior waking time was reported for the cat (Ursin 1971) and dog (Takahashi et al. 1978), and an enhancement of the EEG amplitude in SWS was described for the sleep-deprived rabbit (Pappenheimer et al. 1975). The published data are therefore consistent with the proposition that the close functional relationship between SWS and prior waking constitutes a general feature of mammalian sleep.

3.3 Sleep Homeostasis as a Function of the Circadian Rhythm

The SD experiment illustrated in Fig. 7 was so designed that the onset of recovery sleep coincided with the onset of the light period, the rat's usual sleep-phase. In a further experiment the 24-h SD period was made to terminate before dark onset (Fig. 8). This created a "conflict" between the enhancement of sleep propensity caused by the 24-h vigil, and the tendency for waking and activity as a consequence of the circadian rhythm and the lighting condition. The records of Fig. 8 (right part) show that a sleep rebound occurred in the first dark hours following SD. There was a massive increase of REMS in the dark period (12.0% of recording time vs 3.0% under baseline conditions; $P<0.001$), while NREMS was only slightly increased (24,8% vs 19.5%; $P<0.1$) (Borbély and Neuhaus 1979). SWS exhibited a two-stage compensatory response which consisted of an initial rise during the first dark hours, and a delayed increase during the first light hours (Fig. 8; hatched area). The results demonstrate that sleep and waking continue to exhibit their usual circadian rhythm even after SD. Moreover, the two-stage rebound of SWS indicates that the rapid

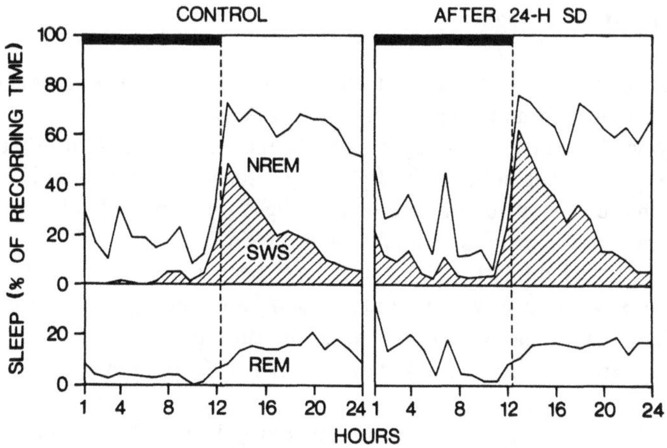

Fig. 8. Effect of sleep deprivation on sleep state pattern in the rat (n=5). Note that, in contrast to **Fig. 7**, recovery sleep starts at dark onset (modified from Borbély and Neuhaus 1979)

compensatory response to a sleep deficit may be impaired if recovery sleep coincides with the circadian phase of waking.

Although the recovery from SD as a function of the circadian rhythm has not yet been investigated in detail in human studies, there is evidence that the situation may be similar to that observed in the rat. Thus, on the one hand the increased sleep propensity of human subjects after one night without sleep was evident from the markedly reduced sleep latency in the morning (Carskadon and Dement 1977). On the other hand, there was a clear persistence of a circadian rhythm of vigilance (e.g. Gulevich et al. 1966) and performance (e.g. Aschoff et al. 1972) during prolonged sleep deprivation. The question whether recovery from sleep loss may be adversely affected by circadian parameters has considerable practical relevance for occupational medicine (e.g. shift-work, time-zone shifts of air-flight personnel) and deserves further investigation.

4 Environmental Variables and Drugs as Tools for Investigating Sleep Regulation

A common approach to the study of sleep regulation consists in a specific modification of the external or internal milieu of the organism. Commonly used environmental variables include sensory stimuli which are varied in intensity to determine the arousal threshold, or ambient temperature which is controlled at various levels to study the relationship between sleep and thermoregulation. Modification of the *milieu intérieur* is often achieved by administering a pharmacological compound which has a selective effect on a neurochemical system, and thereby may reveal links between sleep regulation and specific biochemical processes in the brain. The following discussion centers on two variables which have potent effects on sleep in animals, and may therefore be used as tools for studying sleep regulation.

4.1 Effects of Light

Light has a dual action on sleep. For one, it acts as a zeitgeber for the circadian sleep-wake rhythm and thereby determines its phase-relationship relative to the light-dark cycle. Secondly, short light periods may have a potent direct sleep-promoting or sleep-inhibiting action. The following discussion will deal only with the latter aspect. A more general survey of effects of light on sleep and activity has been published in recent review articles (Borbély 1978, 1980a).

Short light-dark (LD) cycles have been shown to modulate the level of sleep or motor activity in various rodents and carnivores (see for references: Borbély 1978; Table 1). In the rat, sleep is enhanced during the light periods, and depressed during the dark periods of a 2-h LD cycle (LD 1:1) (Fig. 9). A positive functional relationship between sleep induction and light intensity has been demonstrated (Borbély et al. 1975). LD cycles as short as 20 min (LD 10:10 min) are able to modulate sleep (Borbély 1976). It is evident from Fig. 9 that the LD-induced variations of sleep are superimposed on a circadian rhythm. The modulation of the sleep level by short LD cycles is largest during the "active period" of the circadian rhythm (the 12-h sleep period with the lowest average sleep level). This is reflected by a larger circadian amplitude of the sleep minima than of the sleep maxima (Fig. 9, interrupted curves). The sleep-inducing effect of light may be therefore more potent than the sleep-inhibiting effect of darkness, an interpretation that is supported by the higher amount of total sleep during short LD schedules in comparison to LD 12:12 baseline conditions (Borbély et al. 1975; Borbély 1976).

Light does not affect all sleep states of the rat equally. NREMS is readily enhanced by a 1-h light period, whereas REMS is initially inhibited (Borbély et al. 1975; Borbély 1976). Conversely, a selective induction of REMS during the first minutes of a dark period has been repeatedly demonstrated (for references see Borbély 1980a). The latter effect may be used for entraining the rat's intrinsic NREM-REM sleep cycle by a short LD cycle of corresponding period length (approximately 10 min).

Taken together, the experiments illustrate that light, a ubiquitous environmental variable, constitutes a potent experimental tool for sleep research. In the rat, light exerts an intensity-dependent, rapid, sleep-promoting effect with a differential action on the two sleep states. Although the mechanism of action is still obscure, an attempt is made in Sect. 5 to account for the effect of light within the framework of a model.

4.2 Effects of p-Chlorophenylalanine

By inhibiting tryptophan hydroxylase, *p*-chlorophenylalanine (PCPA) causes a long-lasting reduction in brain serotonin (5–HT). PCPA causes a prolonged insomnia in the cat (Delorme et al. 1966; Koella et al. 1968). Administration of 5-hydroxytryptophan leads to a temporary rise in the brain 5-HT level and a temporary restoration of sleep. These results contributed substantially to the establishment of the hypothesis that serotonergic mechanisms play a key role in sleep regulation (see Jouvet 1977 for a recent review). However, the general validity of the concept was questioned on the basis of conflicting results in other species. Thus it was reported for the rat

Fig. 9. Sleep in the rat during a 1-h light/1-h dark schedule. Mean values (30 days recorded in five rats) obtained after phase adjustment of individual records. *Active period* denotes 12-h period with minimum sleep. Dark periods are indicated by *shading*. Mean values of two adjacent light-time peaks and dark-time troughs are connected by *interrupted curves* (modified from Borbély 1975)

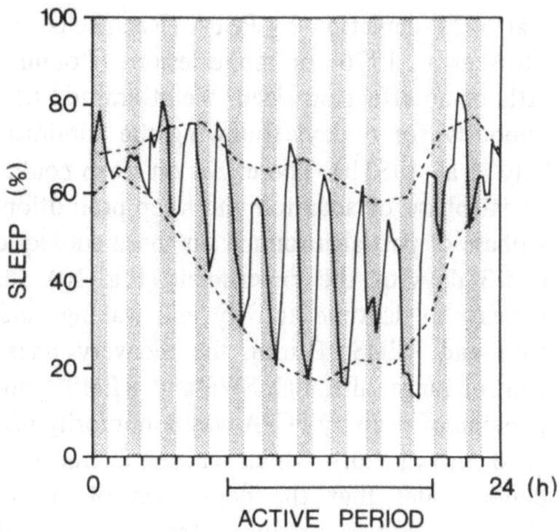

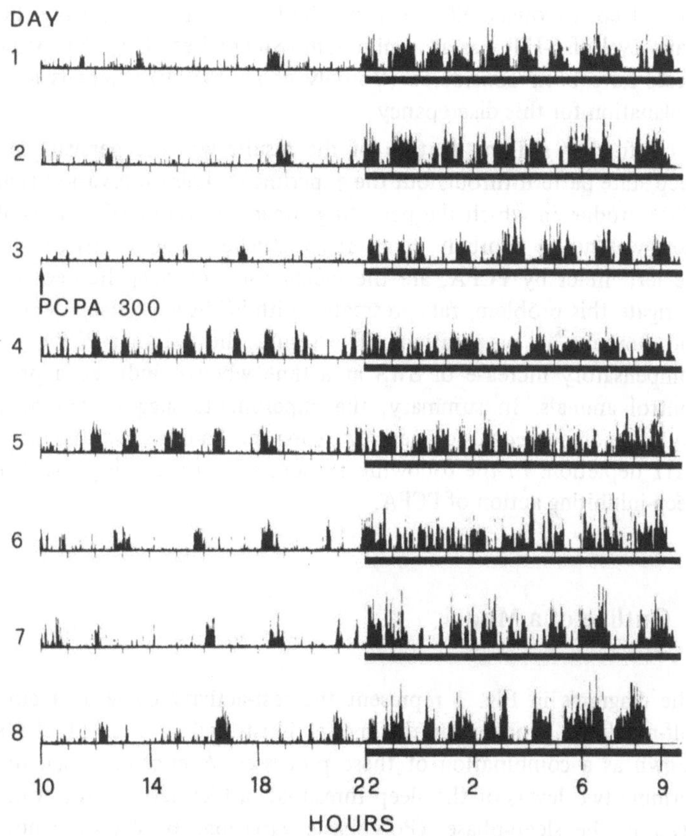

Fig. 10. Effect of *p*-chlorophenylalanine (PCPA 300 mg/kg: day 3, *arrow*) on motor activity in the rat. Motor activity recorded on 2 control days and 6 post-drug days is plotted for successive 10-s epochs in arbitrary units. Dark period (22–10 h) is indicated by *horizontal bar* below records

that 5-HT depletion by PCPA (Rechtschaffen et al. 1973), 5,7-dihydroxytryptamine (Ross et al. 1976) or raphe lesions (Bouhuys and Van den Hoofdakker 1977) had little or no effect on sleep. We attempted to clarify the issue by recording sleep continuously for 6 days following the administration of PCPA (300 mg/kg i.p.) (Borbély et al. 1981b). The effect on sleep could be subdivided into three phases: (1) an initial phase of sedation and sleep-promotion on the day of drug administration; (2) a phase of partial insomnia on the 2 subsequent days; (3) a phase of recovery on the last 3 days of the experiment (Fig. 10). Insomnia was characterized by a marked increase in daytime activity and waking, and by a particularly strong depression of SWS and REMS. During the recovery phase, the values were generally within the control range, although SWS and REMS tended to exceed the baseline level. The depression of sleep by PCPA was temporarily reversed after administration of tryptophan (150 mg/kg i.p.) which causes a transitory increase in brain 5-HT. This result demonstrates that the depression of sleep by PCPA is causally related to 5-HT depletion, and is not due to some non-specific effect. However, there was a striking dissociation between the timecourse of behavioural and neurochemical parameters. Thus sleep no longer differed from the baseline level at a time (PCPA+3d) when the brain level of 5-HT was at a minimum (Steigrad et al. 1978). Although several hypotheses have been considered (Borbély et al. 1981b), there is as yet no satisfactory explanation for this discrepancy.

A further striking feature of the results was the persistence of the usual daily sleep state pattern throughout the experiment. This observation is in line with previous PCPA studies in which the persisting circadian rhythm of various physiological parameters was noted (Borbély et al. 1973; Zucker et al. 1976). If the circadian rhythms are left intact by PCPA, are the mechanisms of sleep homeostasis impaired? To investigate this problem, rats pretreated with PCPA were sleep deprived for 24 h (Tobler and Borbély, in preparation). The results showed that PCPA did not prevent the compensatory increase of SWS at a time when it induced a prominent insomnia in control animals. In summary, the experiments suggest that neither the sleep-wake rhythm nor homeostatic sleep-regulating mechanisms are disrupted by PCPA-induced 5-HT depletion. In the following section, a hypothesis is presented to account for the sleep-inhibiting action of PCPA.

5 Outline of a Model

The diagrams in Fig. 1 represent the rest-activity cycle as a clock-triggered or as a self-regulated process. In the model illustrated in Fig. 11 the sleep-wake cycle is shown as a combination of these processes. A circadian oscillator is assumed to determine two levels of the sleep threshold: a high level in the waking-phase and a low level in the sleep-phase. (Polyphasic variations of sleep are not represented in the diagram.) The self-regulated process is represented by a regulatory variable whose level increases as a function of waking time, and decreases as a function of sleep time. The propensity for sleep or waking is assumed to depend on the level of the regulatory variable relative to the sleep threshold (Fig. 11: hatched areas). During the usual

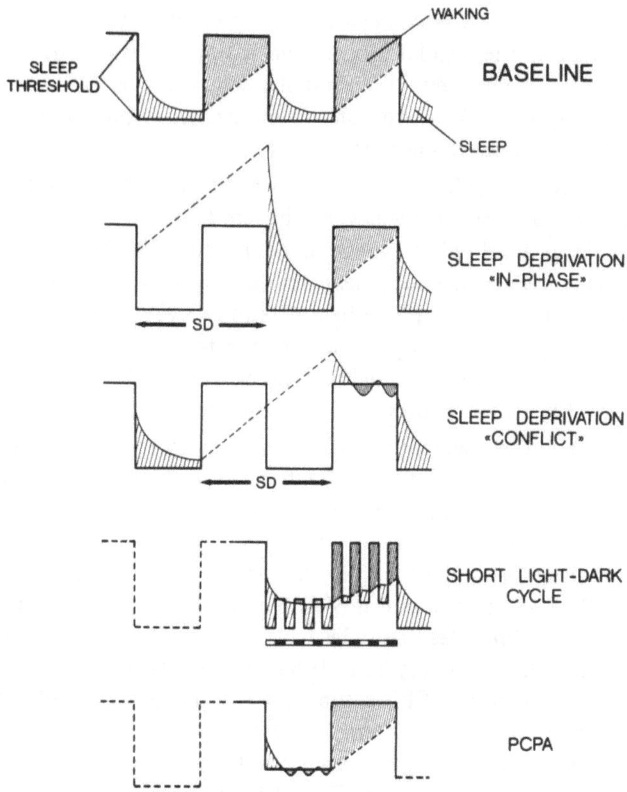

Fig. 11. Model of sleep regulation based on the combination of a circadian oscillator and a self-regulated process

sleep-wake cycle, sleep propensity is highest at the beginning of the daily sleep phase, and declines gradually with the progression of sleep. Since sleep "depth" (intensity) is assumed to correspond to sleep propensity, the widely hatched area reflects also a sleep intensity parameter. As has been pointed out before (Sect. 3), the slow wave sleep fraction of NREMS is a likely indicator for the intensity of a sleep process. The curvilinear decrement of the regulatory variable during sleep corresponds to the decline of SWS or EEG power density in rat and man which may be approximated by an exponential function (Borbély 1981; Borbély et al. 1981 a).

The second and third diagrams of Fig. 11 illustrate the sleep-wake cycle during sleep-deprivation (SD) schedules. The period of forced waking, extending over one sleep-wake cycle, is paralleled by a progressive increase in the level of the regulatory variable. The recovery period coinciding with the usual sleep phase (Fig. 11; "in-phase") therefore shows a high initial sleep intensity which is assumed to be reflected by the high level of SWS (Figs. 6, 7). Due to the curvilinear decline, sleep intensity at the end of the sleep phase is only slightly higher than under baseline conditions. Nevertheless, this difference may be sufficient to cause a residual effect of SD in the subsequent sleep phase (see Fig. 6).

If the end of SD coincides with the onset of the circadian waking phase (Fig. 11; "conflict"), the regulatory variable is only a little above the sleep threshold. Consequently, a relatively short initial sleep period is followed by a period of frequent waking episodes. However, due to the persisting high level of the regulatory variable, a delayed increase in sleep intensity is observed during the circadian sleep phase (Fig. 8).

Short LD cycles produce a powerful modulation of the sleep level, but do not disrupt the circadian sleep-wake rhythm (Fig. 9). In the model illustrated in Fig. 11, the sleep threshold is shown to be affected by the lighting conditions. During the circadian sleep phase, darkness facilitates waking by raising the threshold, while during the circadian waking phase light promotes sleep by lowering the threshold. The sleep-promoting action of light is larger than the converse effect of darkness. This feature as well as the small LD-induced modulation at the beginning of the circadian sleep phase (Fig. 9) are incorporated in the model.

The lowest diagram in Fig. 11 is designed to simulate the effect of PCPA which in the rat causes a reduction of daytime sleep. The drug is assumed to act by raising the daytime sleep threshold while leaving other aspects of sleep regulation unchanged. As a consequence, sleep occurs in the very first part of the circadian sleep phase and intermittently later on, while the waking phase remains practically unaffected (Fig. 10). Both the restriction of SWS to the first part of the light phase, and the shortening of light-time sleep episodes have been documented for the 1 st day following PCPA administration (Borbély et al. 1981 b). The assumption that sleep regulation as a whole is not disrupted by PCPA is supported by the invariance of the daily sleep state pattern and the SWS rebound after SD (see Sect. 4). A similar interpretation of the effects of 5-HT depletion has been advanced by Trulson and Jacobs (1979) and Ursin (1980).

The model presented in Fig. 11 is based on several simplifications and assumptions which are, however, not crucial for its main features. They include the square wave function of the sleep threshold which could be equally represented by a sine wave. Furthermore, it could be postulated that the short LD cycles and PCPA affect the regulatory variable and not the sleep threshold. Since the important variable for the model is the difference between these two parameters, this alternative assumption would not cause a basic change. Another simplification relates to the sleep cycles in men and the polyphasic sleep pattern in animals (e.g. the rat) which are not represented in Fig. 11, but could be easily incorporated into the model.

Finally, it should be pointed out that throughout this review the focus has been on NREM sleep, whereas REM sleep has been only marginally considered. The major reason for this bias has been the homeostatic regulation of NREMS which is so clearly reflected by EEG parameters, while REMS regulation is less easily characterized (see Borbély and Neuhaus 1979; and Borbély 1980b for further discussions of this aspect). A few words are therefore justified to indicate how this important substate of sleep could be dealt with in a more elaborate version of the model.

On a two-dimensional state-space whose axes are defined by the EEG arousal level (a_E, A_E) and the behavioural arousal level (a_B, A_B) (a and A denoting low and high arousal, respectively), waking may be assumed to corresponds to A_E, A_B; NREMS to a_E, a_B; and REMS to A_E, a_B. REMS, exhibiting aspects of EEG arousal together with behavioural sleep, may be viewed as a "metastable" sleep state which occurs when the regulatory variable falls slightly below the sleep threshold (Fig. 11), and which could serve to maintain the sleep process during such periods. According to this hypothesis, REMS would make it possible for the brain to accommodate intrinsic or extrinsic arousal stimuli (reflected by EEG arousal) without disruption of sleep. Thus the NREM–REM sleep cycle may result from periodic intrinsic arousal processes, while REMS induction by dark periods in the rat (Sect. 4) may be considered to be the effect of an extrinsic arousal parameter. The model predicts that the probability of REMS occurrence increases towards the end of the sleep phase as the regulatory variable approaches the sleep threshold, a prediction borne out by data in rat and man (Borbély 1980a). However, further assumptions are warranted to account for the homeostatic aspects of REMS regulation.

The model outlined in this section is based on the combination of the two regulatory mechanisms illustrated in Fig. 1. The circadian oscillator allows sleeping and waking to occur at predictable intervals in synchrony with the environmental day-night cycle. However, the very stability of the circadian rhythm (Figs. 3, 4) may impair the adjustment of the organism to unexpected challenges from the environment. A self-regulated process seems to be more apt to cope with this requirement, since it renders sleep contingent on the duration of waking. On the other hand, synchrony with the environmental cycle is not maintained by such a process. The combination of the two mechanisms may therefore provide an optimal solution to the requirements of stability and flexibility.

The two types of regulatory processes proposed in this article are based on a phenomenological approach to sleep regulation which leaves the underlying neurobiological mechanisms largely unspecified. This may be an advantage insofar as major functional aspects may be analysed without undue regard to current neurobiological notions and theories. On the other hand, the model of sleep homeostasis rests heavily on EEG parameters whose generating mechanisms and origins are obscure. It is therefore desirable to validate the concept on the basis of other measurable parameters. Such a possibility may arise from recent developments indicating that sleep propensity in animals and man may be regulated by a humoral factor (Fencl et al. 1971; Nagasaki et al. 1974; Borbély and Tobler 1980; Krueger et al. 1980). Particularly relevant to the present arguments are reports that the CSF concentration of a still unidentified sleep factor is a function of prior waking time (Fencl et al. 1971), and that the factor induces the same type of high-amplitude slow wave sleep as observed after sleep deprivation (Pappenheimer et al. 1975). These results show that the investigation of sleep homeostasis may be successfully expanded to include neurochemical aspects. Moreover, there is the intriguing possibility that the regulatory variable of the present model may turn out to be an endogenous sleep-regulating compound.

6 Summary

Two facets of sleep are relevant for sleep regulation: (1) the circadian sleep-wake rhythm which is controlled by a highly stable circadian pacemaker; and (2) sleep homeostasis which is reflected by the constant average level of sleep and by its compensatory increase following a sleep deficit. Loss of non-REM sleep may be compensated for not only by a prolongation of sleep time, but also by an increase in apparent sleep intensity. Slow wave sleep seems to represent the high-intensity fraction of non-REM sleep. This assumption is based on results obtained by frequency analysis of animal and human EEG which indicate that the proportion of EEG slow waves is determined by the duration of the preceding waking period. A conflict may arise between the circadian and the homeostatic sleep regulating mechanism, if the period of sleep compensation is made to coincide with the circadian phase of minimum sleep. In this situation, rats partly delay the compensation of a non-REM sleep deficit to the circadian phase of maximum sleep. Homeostatic regulation depends therefore on the phase of the circadian rhythm in which sleep occurs. The circadian and the homeostatic aspects of sleep are combined in a model of sleep regulation. It is assumed that the sleep threshold is controlled by a circadian oscillator, while an additional regulatory variable is determined by the prior vigilance state. Sleep propensity or sleep intensity is represented in the model by the difference between the level of the regulatory variable and the sleep threshold. Effects of sleep deprivation, short light-dark cycles and p-chlorophenylalanine (a serotonin synthesis inhibitor) on rat sleep are discussed on the basis of experimental and simulated data.

Acknowledgements. I thank Dr. Irene Tobler for her constructive criticism of the manuscript. The help from Ms. K. Schwarz (preparation of figures) and from Ms. R. Steiger (typing) is gratefully acknowledged. The experimental work reported in this paper was supported by the Swiss National Science Foundation, grants. nr. 3.254-0.77 and 3.561.-0.79.

References

Agnew HW Jr, Webb WB, Williams RL (1964) The effect of stage four sleep deprivation. Electroencephalogr Clin Neurophysiol 17:68–70

Aschoff J, Giedke H, Poppel F, Wever R (1972) The influence of sleep-interruption and of sleep-deprivation on circadian rhythms in human performance. In: Colquhoun WP (ed) Aspects of human efficiency. The English University Press, London, pp 135–150

Berger RJ, Oswald I (1962) Effects of sleep deprivation on behaviour, subsequent sleep and dreaming. J Ment Sci 108:457–465

Blake H, Gerard RW (1937) Brain potentials during sleep. Am J Physiol 119:692–703

Borbély AA (1975) Circadian rhythm of vigilance in rat: modulation by short light-dark cycles. Neurosci Lett 1:67–71

Borbély AA (1976) Sleep and motor activity of the rat during ultrashort light-dark cycles. Brain Res 114:305–317

Borbély AA (1977) Sleep in the rat during food deprivation and subsequent restitution of food. Brain Res 124:457–471

Borbély AA (1978) Effects of light on sleep and activity rhythms. Prog Neurobiol 10:1–31

Borbély AA (1980a) Effects of light and circadian rhythm on the occurrence of REM-sleep in the rat. Sleep 2:289–298

Borbély AA (1980b) Sleep: circadian rhythm versus recovery process. In: Koukkou M, Lehmann D, Angst J (eds) Functional states of the brain: their determinants. Elsevier, Amsterdam, pp 151–161

Borbély AA (1981) The sleep process: circadian and homeostatic aspects. In: Obál F, Benedek G (eds) Environmental physiology. Adv Physiol Sciences 18:85–91

Borbély AA, Neuhaus HU (1978a) Daily pattern of sleep, motor activity and feeding in the rat: effects of regular and gradually extended photoperiods. J Comp Physiol 124:1–14

Borbély AA, Neuhaus HU (1978b) Circadian rhythm of sleep and motor activity in the rat during skeleton photoperiod, continuous darkness, and continuous light. J Comp Physiol 128:37–46

Borbély AA, Neuhaus HU (1979) Sleep-deprivation: effects on sleep and EEG in the rat. J Comp Physiol 133:71–87

Borbély AA, Tobler I (1980) The search for an endogenous "sleep-substance". Trends in Pharmacology 1:356–358

Borbély AA, Huston JP, Waser PG (1973) Physiological and behavioral effects of parachlorophenylalanine in the rat. Psychopharmacology (Berlin) 31:131–142

Borbély AA, Huston JP, Waser PG (1975) Control of sleep states in the rat by short light-dark cycles. Brain Res 95:89–101

Borbély AA, Neuhaus HU, Tobler I (1981b) Effect of p-chlorophenylalanine and tryptophan on sleep, EEG and motor activity in the rat. Behav Brain Res 2:1–22

Borbély AA, Baumann F, Brandeis D, Strauch I, Lehmann D (1981a) Sleep-deprivation: effect on sleep stages and EEG power density in man. Electroencephalogr Clin Neurophysiol 51:483–493

Bouhuys AL, Van den Hoofdakker RH (1977) Effects of midbrain raphe destruction on sleep and locomotor activity in rats. Physiol Behav 19:535–541

Cannon WB (1939) The wisdom of the body. Norton, New York

Carskadon MA, Dement WC (1977) Sleep tendency: an objective measure of sleep loss. In: Chase MH, Mitler MM, Walter PL (eds) Sleep research, vol 6. Brain Research Institute, Los Angeles, p 200

Church MW, March JD, Hibi S, Cavness C, Feinberg I (1975) Changes in frequency and amplitude of delta activity during sleep. Electroencephalogr Clin Neurophysiol 39:1–7

Crowley TJ, Kripke DF, Halberg F, Pegram GV, Schildkraut JJ (1972) Circadian rhythms of Macaca mulatta: sleep, EEG, body and eye movement, and temperature. Primates 13:149–168

Danguir J, Nicolaidis S (1980) Circadian sleep and feeding patterns in the rat: possible dependence on lipogenesis and lipolysis. Am J Physiol 238:E223–E230

Delorme F, Froment JL, Jouvet M (1966) Suppression du sommeil par la p. chlorométhamphétamine et la p. chlorophénylalanine. C R Soc Biol (Paris) 160:2347–2351

Dement W (1960) The effects of dream deprivation. Science 131:1705–1707

Eastman CI (1980) Circadian rhythms of temperature, waking, and activity in the rat: dissociations, desynchronizations, and disintegrations. PhD thesis, University of Chicago

Feinberg I, Floyd TC (1979) Systematic trends across the night in human sleep cycles. Psychophysiology 16:283–291

Feinberg I, Fein G, Walker JM, Price LJ, Floyd TC, March JD (1979) Flurazepam effects on sleep EEG. Arch Gen Psychiatry 36:95–102

Fencl V, Koski G, Pappenheimer JR (1971) Factors in cerebrospinal fluid from goats that affect sleep and activity in rats. J Physiol (Lond) 216:565–589

Friedman L, Bergmann BM, Rechtschaffen A (1979) Effects of sleep deprivation on sleepiness, sleep intensity, and subsequent sleep in the rat. Sleep 1:369–391

Gulevich G, Dement W, Johnson L (1966) Psychiatric and EEG observations on a case of prolonged (264 h) wakefulness. Arch Gen Psychiatry 15:29–35

Hess WR (1965) Sleep as a phenomenon of the integral organism. Prog Brain Res 18:3–7

Ibuka N, Kawamura H (1975) Loss of circadian rhythm in sleep-wakefulness cycle in the rat by suprachiasmatic nucleus lesions. Brain Res 96:76–81

Ibuka N, Inouyé ST, Kawamura H (1977) Analysis of sleep-wakefulness rhythms in male rats after suprachiasmatic nucleus lesions and ocular enucleation. Brain Res 122:33–47

Jouvet M (1977) Neuropharmacology of the sleep-waking cycle. In: Iversen SD, Iversen LL, Snyder SH (eds) Handbook of psychopharmacology, vol 8. Plenum, New York, pp 233–293

Karacan I, Williams RL, Finley WW, Hursch CJ (1970) The effect of naps on nocturnal sleep: influence on the need for stage-1 REM and stage-4 sleep. Biol Psychiatry 2:391–399

Koella WP, Feldstein A, Czicman JS (1968) The effect of parachlorophenylalanine on the sleep of cats. Electroencephalogr Clin Neurophysiol 25:481–490

Krueger JM, Bacsik J, Garcia-Arraras J (1980) Sleep-promoting material from human urine and its relation to factor S from brain. Am J Physiol 283:E116–E123

Le Magnen J, Tallon S (1968) L'effet du jeune préalable sur les caractéristiques temporelles de la prise d'aliments chez le rat. J Physiol (Paris) 58:143–154

Morden B, Mitchell G, Dement W (1967) Selective REM sleep deprivation and compensation phenomena in the rat. Brain Res 5:339–349

Moses JM, Johnson LC, Naitoh P, Lubin A (1975) Sleep stage deprivation and total sleep loss: effects on sleep behavior. Psychophysiology 12:141:146

Nagasaki H, Iriki M, Inoué S, Uchizono K (1974) The presence of a sleep-promoting material in the brain of sleep-deprived rats. Proc Jpn Acad 50:241–247

Nakazawa Y, Kotorii M, Ohishima M, Kotorii T, Hasuzawa H (1978) Changes in sleep pattern after sleep deprivation. Folia Psychiatr Neurol Jpn 32:85–93

Pappenheimer JR, Koski G, Fencl V, Karnovsky ML, Krueger J (1975) Extraction of sleep-promoting factor S from cerebrospinal fluid and from brains of sleep-deprived animals. J Neurophysiol 38:1299–1311

Pittendrigh CS, Daan S (1976) A functional analysis of circadian pacemakers in nocturnal rodents. J Comp Physiol 106:223–252

Rechtschaffen A, Kales A (eds) (1968) A manual of standardized terminology, techniques, and scoring system for sleep stages of human subjects. National Institutes of Health Publication 204. US Government Printing Office, Washington, DC

Rechtschaffen A, Lovell RA, Freedman DX, Whitehead WE, Aldrich M (1973) The effect of parachlorophenylalanine on sleep in the rat: some implications for the serotonin-sleep hypothesis. In: Barchas J, Usdin E (eds) Serotonin and behavior. Academic Press, New York, pp 401–424

Reite ML, Rhodes JM, Kavan E, Adey WR (1965) Normal sleep patterns in macaque monkey. Arch Neurol 12:133–144

Rosenberg RS, Bergmann BM, Rechtschaffen A (1976) Variations in slow wave activity during sleep in the rat. Physiol Behav 17:931–938

Ross CA, Trulson ME, Jacobs BL (1976) Depletion of brain serotonin following intraventricular 5,7-dihydroxytryptamine fails to disrupt sleep in the rat. Brain Res 114:517–523

Steigrad P, Tobler I, Waser PG, Borbély AA (1978) Effect of p-chlorophenylalanine on cerebral serotonin binding, serotonin concentration, and motor activity in the rat. Naunyn-Schmiedebergs Arch Pharmacol 305:143–148

Takahashi Y, Ebihara S, Nakamura Y, Takahashi K (1978) Temporal distributions of delta wave sleep and REM sleep during recovery sleep after 12-h forced wakefulness in dogs; similarity to human sleep. Neurosci Lett 10:329–334

Trulson ME, Jacobs BL (1979) Raphe unit activity in freely moving cats: correlation with level of behavioral arousal. Brain Res 163:135–150

Ursin R (1971) Differential effect of sleep deprivation on the two slow wave sleep stages in the cat. Acta Physiol Scand 83:352–361

Ursin R (1981) Does para-chlorophenylalanine produce disturbed waking, disturbed sleep or activation by ponto-geniculo-occipital waves in cats? Waking Sleeping 4:211–221

Webb WB, Agnew HW Jr (1971) Stage 4 sleep: influence of time course variables. Science 174:1354–1356

Webb WB, Agnew HW Jr (1974) The effects of a chronic limitation of sleep length. Psychophysiology 11:265–274

Weitzman ED, Czeisler CA, Moore-Ede MC (1979) Sleep-wake, neuroendocrine and body temperature circadian rhythms under entrained and non-entrained (free-running) conditions in man. In: Suda M, Hayaishi O, Nakagawa H (eds) Biological rhythms and their central mechanism. Elsevier, Amsterdam, pp 199–227

Wever RA (1979) The circadian system of man. Springer, Berlin Heidelberg New York

Williams HL, Hammack JT, Daly RL, Dement WC, Lubin A (1964) Responses to auditory stimulation, sleep loss and the EEG stages of sleep. Electroencephalogr Clin Neurophysiol 16:269–279

Zucker I, Rusak B, King RC Jr (1976) Neural bases for circadian rhythms in rodent behavior. Adv Psychobiol 3:35–74

Haemodynamic Changes During Sleep in Man*

John V. Jones[1], Peter Sleight[1], and Harley S. Smyth[2]

Contents

Abbreviations

BPM: beats per minute; EEG: electroencephalogram; EOG: electrooculogram; REM: rapid eye movement

1 Introduction

Relatively little is known about the haemodynamic changes which accompany sleep in man. Sleep itself is not homogeneous and it is hardly surprising that conflicting results have been obtained from studies carried out in the past. Newer techniques, such as those involving ambulatory monitoring or electroencephalography, are gradually helping to resolve some of the difficulties.

---

* Supplemented by grants from the Medical Research Council and the British Heart Foundation

1 Department of Cardiovascular Medicine, University of Oxford, John Radcliffe Hospital, Oxford, Ocap3 9DV, Great Britain

2 Present address: Division of Neurosurgery, Toronto General Hospital, Toronto, Canada

It has been known since the time of Galen that sleep is of haemodynamic significance. He reported that the "sleeping pulse of man is much depressed." Numerous investigators have since confirmed this observation and have extended it to include other circulatory parameters. Blood pressure, for instance, was also shown to fall with sleep in dogs as long ago as 1894, a fact soon confirmed in man by Howell (1897) who found dilatation of some vessels and a fall in blood pressure in sleeping subjects.

In hypertensive subjects the fall in blood pressure with sleep may be greater than in normotensive people. Somewhat paradoxically, however, in severe or malignant phase hypertension the blood pressure fall with sleep may be greatly diminished or even abolished (Shaw et al. 1963). Similarly, with phaeochromocytoma blood pressure may fall less with sleep (Littler and Honour 1974) than in other types of hypertension of similar severity.

Sudden acute rises in pressure occurring during sleep were first noted by MacWilliam in 1923. He suggested that dreams might be involved in the generation of such blood pressure rises. Others have since attempted to correlate haemodynamic events with changes in the electroencephalogram. Two studies published in 1964 noted a relationship between the depth of sleep determined by the EEG and the fall in blood pressure (Snyder et al. 1964; Richardson et al. 1964).

There had been some controversy as to whether recumbency could account in toto for the reduction in blood pressure or whether sleep itself was a necessary prerequisite for this event (Smirk 1957). Richardson et al. (1965) thought that 60% of the fall in arterial pressure seen at night time was in fact dependent upon sleep itself.

Few studies have been performed to determine the relative roles of cardiac output and peripheral resistance in causing the fall in sleeping blood pressure. Those that have been performed (Khatri and Freis 1967a, b; Bristow et al. 1969) have given conflicting results.

We intend to review the studies performed in our laboratory on various haemodynamic aspects of sleep and relate them to the work of others in this field.

2 Methods

Subjects were studied on three successive nights with gradually increasing complexity of instrumentation. The EEG was measured each night using five channels arranged unilaterally, together with electrodes placed behind the angle of each eye (to record the EOG) and submentally (to record activity in axial muscle, platysma). Respiration was measured by a pneumogram/strain gauge. On the second night cuff blood pressures were obtained at automatic intervals of 2.5 or 5 min using an oscillometric method.

On the third night blood pressure was measured by an indwelling catheter inserted by the micro Seldinger technique into the brachial artery. A catheter was also placed in an antecubital vein for the injection of vaso-active drugs (see below). These cannulae were connected to pressure gauges or syringes via 1 m lengths of polythene tubing so as to allow freedom of movement during sleep. Perhaps rather surprisingly the subject had little difficulty in sleeping despite this instrumentation.

3 Classification of Sleep Stages

Much of the investigation carried out in sleep is of relevance only if the accompanying EEG events are known. As will be seen later, for instance, even the simple manoeuvre of inflating a blood pressure cuff can lead to arousal or indeed, as some of our subjects reported, actually prevent the onset of sleep. The classification of the sleep EEG that has been employed in our laboratories is that of Gastaut and Broughton (1965) and is as follow:

Stage Ia. This corresponds clinically to drowsiness. The alpha rhythm increases in amplitude, and becomes slower (normally 8–10/s). It also passes over the scalp to more anterior parts of the head.

Stage Ib. Is very light sleep. The alpha rhythm diminishes and disappears, being replaced by slower, low voltage theta rhythm (4–7/s). There may also be brief bursts of beta activity (30/s) and an occasional vertex short wave of low voltage may follow a sensory stimulus. If such a stimulus (such as noise or light or touch) is sufficiently intense, the alpha rhythm may reappear and the subject is returned to stage Ia.

Stage II. This is the clinical equivalent of light sleep. The predominant rhythm is still of theta frequency but in the central regions there are frequent bursts of 12–14/s activity (sleep spindles). Also occurring quite characteristically at this stage are the "K-complexes." These are large sharp waves spreading from the vertex and followed by a slow wave. K-complexes may be evoked by various sensory stimuli, or they may occur spontaneously without obvious external stimulus.

Stage III. Is moderately deep sleep. The rhythm on the EEG is delta (1–3/s). They are frontal at first and occupy up to 50% of the trace.

Stage IV. This is deep sleep and it can be quite difficult to rouse the individual from it. The delta activity is predominant over the whole scalp.

REM Sleep. The voltage of the EEG is this stage is very low and there is an almost complete absence of reactive EEG transients. The onset is usually abrupt and associated with bursts of rapid eye movements (hence REM). The very low voltage pattern may be interrupted in vertical rhythms by bursts of a rhythm in the theta frequency. The subjects are extremely unreactive in REM sleep but on awakening about 80%–90% can recell having dreams (Dement and Kleitman 1957) and so this stage is often referred to as "dreaming sleep."

4 First-Night Effect

In all sleep studies consideration must be given to the "first-night effect." This, as most will know from personal experience, is manifest as a restless disturbed night of sleep usually seen in the first night in a new environment such as a hospital or hotel.

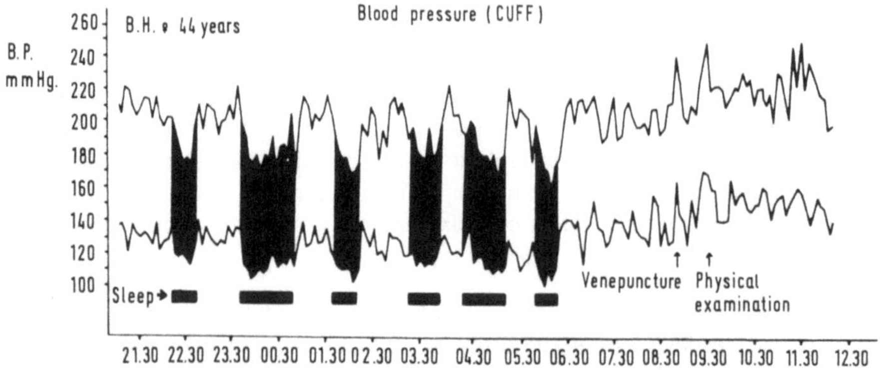

Fig. 1. Patient with high blood pressure. First-night study. Plotted record of blood pressure readings obtained by automatic cuff every 5 min throughout the first night and morning after admission to hospital. Times of sleeping (denoted in *black*) and waking. Note the inability to sleep consistently on this first night and the effect that this has on the blood pressure

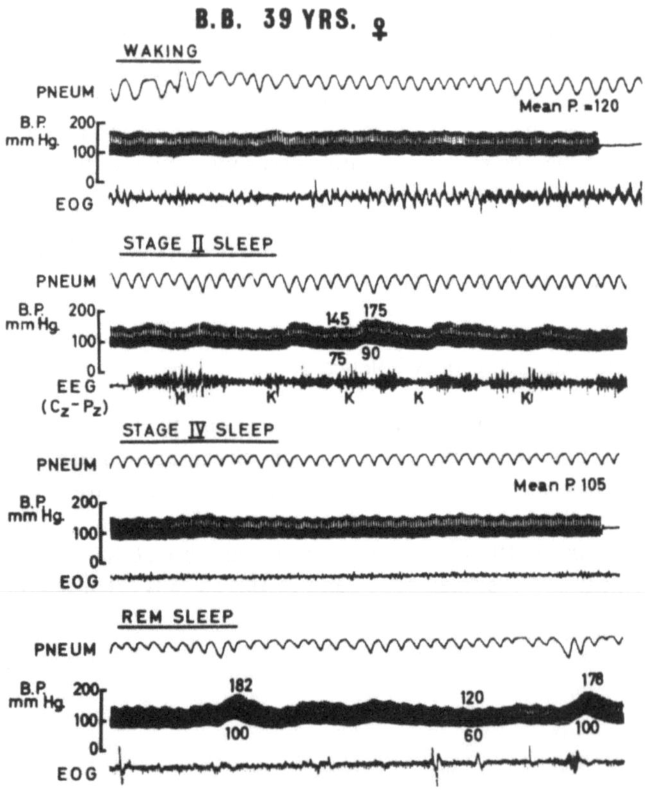

Fig. 2. Serial records of respiration, intra-arterial blood pressure, and eye movements awake and in Stages II, IV, and *REM* sleep. (*EEG*, electroencephalogram, *EOG*, electro-oculogram)

It consists of more and longer periods of wakefulness (Fig. 1), shorter and fewer REM periods and more frequent changes in the EEG stages of sleep. These unwanted environmental effects generally have disappeared by the second and subsequent nights when the subject has settled into his new surroundings. The first-night effect can, therefore, be seen to be of profound importance when studying haemodynamics in the laboratory. This seems to be of less importance when blood pressure is being determined using ambulatory monitoring techniques as will be described later. Here, where the subject is at home to sleep, the pattern of blood pressure throughout the night is very similar for the first and second nights.

5 Neural Control of the Circulation in Sleep

Smyth (1967) found that there was a good correlation between the EEG stages and the pattern of blood pressure recorded (see Fig. 2) during sleep. Indeed, he found the correlation so good that he could predict what type and depth of sleep was in progress when the pressures were recorded. In the recent study of Maling et al. (1979) they too found a very good correlation between stage of sleep on the EEG and the fall in blood pressure. It can be seen from Fig. 2 that all the sleep blood pressures are lower than the awake resting value.

The blood pressure in Stage I will be considered later but Stage II sleep is seen to be less stable with marked fluctuations compared to the awake blood pressure and also compared to Stage IV sleep which again is relatively stable. REM sleep is extremely variable but it too will be considered in greater detail later.

6 Sleep and Arousal and Their Effect on Blood Pressure

It is of interest that minor stimuli can lead from Stage Ib sleep back to Stage Ia. Such stimuli include noise and touch and generally the procedure of blood pressure measurement by cuff includes both such stimuli. For this reason studies where frequent cuff blood pressure measurements have been used to determine haemodynamic events must be viewed with a certain reserve. Figure 3 is a record of blood pressure recorded with the cuff on a subject who subsequently claimed that the onset of his sleep was repeatedly interrupted by the inflation of the cuff. Presumably he was repeatedly being aroused from Stage Ib to Ia by the procedure. The final picture is one of a gradual decline of blood pressure over the early hours of sleep. This may, however, be artefactual as can be seen in comparison with the blood pressure traces of Fig. 4 where pressure was recorded continuously by means of an indwelling arterial catheter in the brachial artery. In these direct records blood pressure falls abruptly at the beginning of sleep and does not have the gradual decline which occurs with cuff recordings. Spikes of arousal are seen at 1.00 and 4.30 a.m.

This point can be made in a different way. Figure 5 shows the effect on intra-arterially recorded blood pressure of inflating a cuff on the other arm. Many cuff

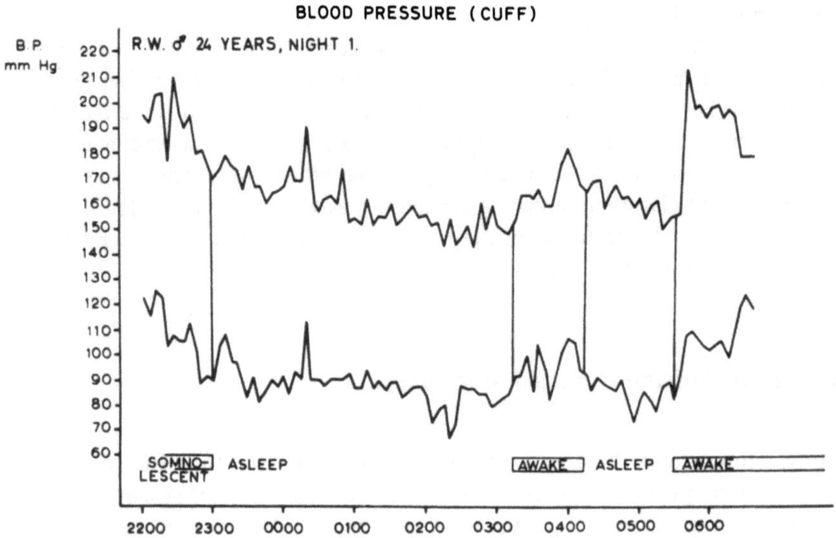

Fig. 3. Night blood pressure study using an automatic cuff oscillometer to measure blood pressure. Note gradual fall in pressure while falling asleep but more abrupt rise on arousal

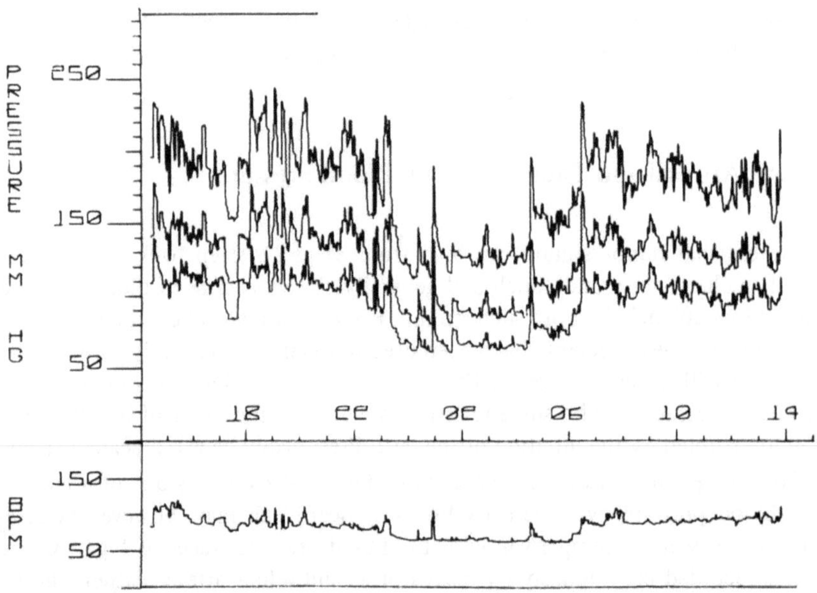

Fig. 4. Twenty-four h intra-arterial ambulatory, blood pressure in a hypertensive patient. Note the abrupt fall in blood pressure that occurs on falling asleep at about 1.00 a.m. This patient rose twice during the night to pass urine and after the second such occasion described his sleep as light and restless. *BPM*, beats per minute, i.e. heart rate

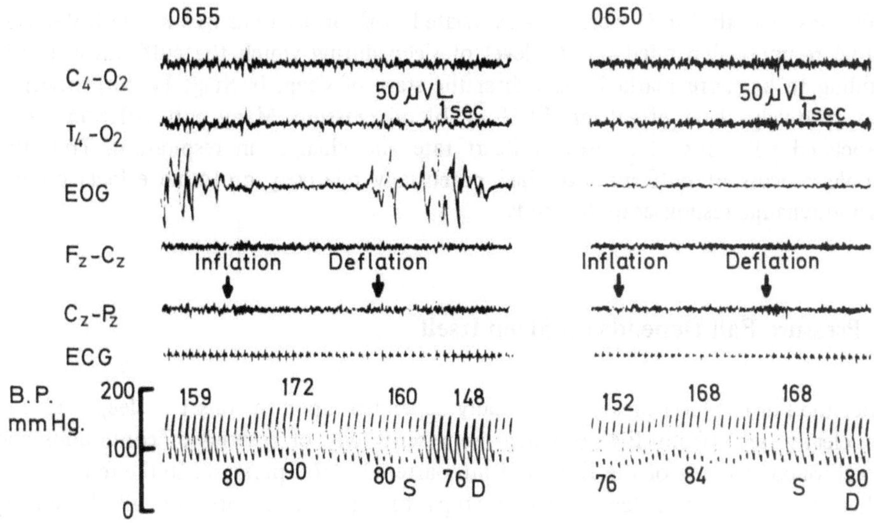

Fig. 5. Effect of cuff inflation upon intra-arterial recorded blood pressure. Left hand example shows a transient effect on the blood pressure but the right hand example shows that inflating the cuff has resulted in a sustained rise in arterial pressure

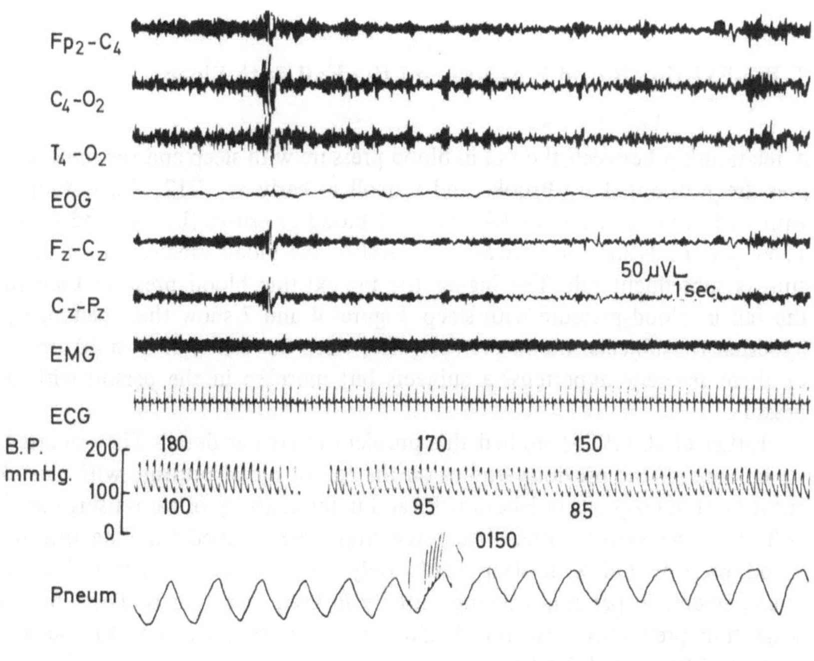

Fig. 6. Rapid transition from Stage Ia to Stage II sleep. There is sudden waxing of the alpha rhythm followed by its disappearance. Arterial pressure falls simultaneously from 180/100 to 150/85

inflations (Smyth 1967) were also associated with arousal changes in the EEG. The actual response depended on the level of sleep during which the cuff was inflated, tending to be more marked the lighter the stage of sleep. In Stage IV cuff inflation had relatively little effect on EEG or blood pressure. Many cuff inflations were associated with brief increases in heart rate and changes in respiration. Thus the simple process of cuff inflation had potentially relatively profound effects on the haemodynamic responses under study.

7 Pressure Fall Depends on Sleep Itself

That the fall in blood pressure is actually dependent upon the onset of sleep itself can be clearly seen from Fig. 6 which represents one of a series of continuous and synchronous records of the EEG and intra-arterial blood pressure. In the few seconds when the alpha region slows and then stops, the mean blood pressure falls 30 mm Hg while the heart rate also slows. As noted earlier, blood pressure levels fall even further as sleep deepens but this initial fall in pressure tends to be steeper.

The opposite pattern occurs with arousal. When the alpha rhythm reappears on the EEG the blood pressure rises rapidly. These events are so quick and so abrupt that they are highly suggestive of neural rather than humoral pressure control mechanisms being involved.

8 Pre-Existing Blood Pressure and the Fall with Sleep

A relationship between the fall in blood pressure with sleep and the pre-existing blood pressure was noted by Brooks and Carroll as early as 1912. They studied subjects with high, low, and moderately elevated blood pressures (Brooks and Carroll 1912). There was a relationship between the level of the blood pressure at the onset of sleep and its subsequent fall. The higher the pre-existing blood pressure then the greater the fall in blood pressure with sleep. Figures 4 and 7 show this phenomenon in two hypertensive subjects. Blood pressure after sleep has fallen at least 40 mm Hg in both of these severely hypertensive subjects but more so in the person with the higher pressure.

Littler et al. (1975) studied this problem in greater detail. They pointed out that all but one of the previous studies on the fall of blood pressure with sleep had been carried out in hospitals or laboratories and in the majority of cases using blood pressure cuffs with necessarily intermittent readings. They argued that the true measure of blood pressure fall with sleep could only be obtained by direct arterial pressure measurements in patients sleeping soundly in their own homes. There had been some suggestion previously that this might make a large difference to the observations recorded (Bevan et al. 1969).

Eighteen patients were studied who had reported in their records that they slept soundly and comfortably. As the patients were at home and not under direct ob-

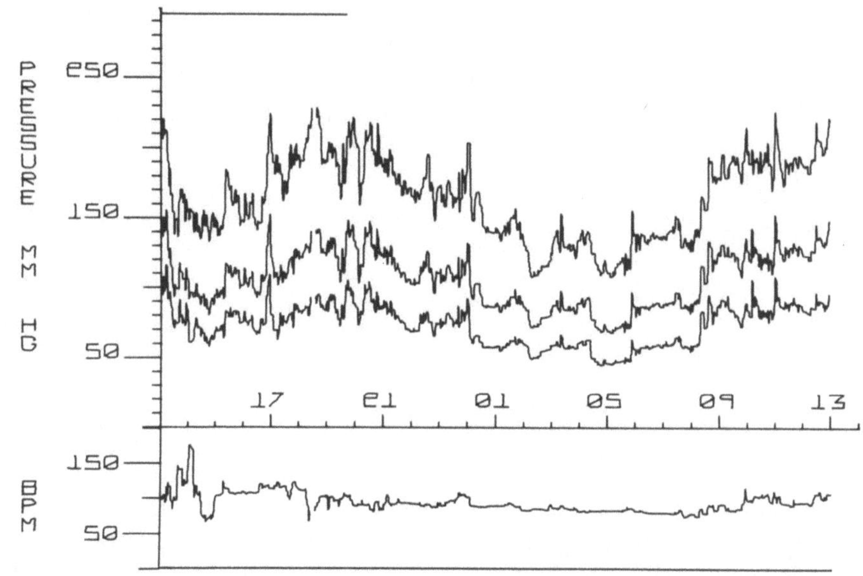

Fig. 7. For comparison with Fig. 4. In this less severely hypertensive subject the blood pressure falls less with sleep when compared to the subject in Fig. 4

servation and also since no EEG was recorded, the period of sleep was taken from the point at which the subject indicated that he was in bed until the time he indicated that he was awake. Eight of these patients were hypertensive and already on a variety of anti-hypertensive medications. Under these circumstances blood pressure fell about 20% in all the subjects studied. This is when the whole of the waking blood pressure averaged was compared with the sleeping pressure also averaged. It is evident that a 20% fall will be greater in real values in a hypertensive person compared to normotensive individuals.

They noted, to their surprise, that treatment with neuronal blocking agents with which some of the patients were being treated did not alter the blood pressure reduction with sleep although no comparison was made in the same subjects before and after treatment.

9 Mechanism of the Blood Pressure Fall with Sleep

Exactly what causes the fall in blood pressure with sleep is unknown. Although, as will be discussed later, there are circadian rhythms of such substances as catecholamines and other humoral agents, these clearly cannot explain the abrupt reduction in pressure that seems to occur with the loss of the alpha rhythm from the EEG. In sleep there seems to be a trend towards dominant parasympathetic activity with a reduction in sympathetic tone. This is manifest by the bradycardia and miosis of sleep.

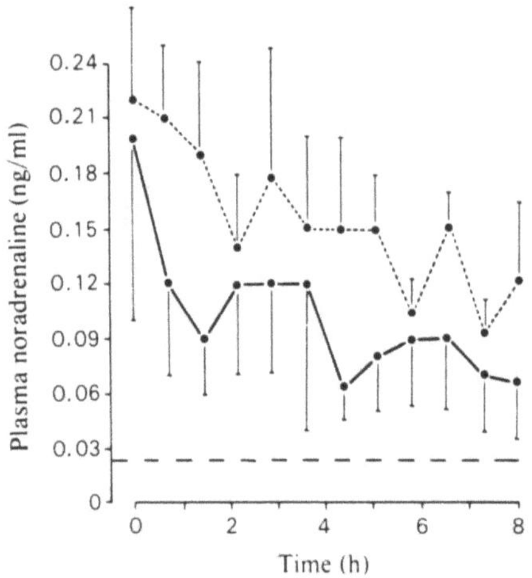

Fig. 8. Mean plasma noradrenaline concentration for subjects on placebo (●– – – –●) or clonidine (●———●). Note the steady fall in plasma noradrenaline with length of time asleep. (Maling et al. 1979)

Few studies have been made of the effect of sleep on the cardiac output, Khatri and Freis (1967a, b) were the first to perform such studies in man. They found that the cardiac output fell about 10% in normotensive subjects but the peripheral resistance remained unchanged except in hypertensive subjects where some fall did occur with sleep. A somewhat more detailed haemodynamic study was performed by Bristow et al. (1969) who arbitrarily divided their subjects into normotensive (mean blood pressure less than 100 mm Hg) or hypertensive (mean blood pressure greater than 100 mm Hg). Resting cardiac outputs awake were the same for both groups but no change in cardiac output resulted from sleep in the normotensive subjects while there was a slight fall in the hypertensive group although this did not achieve statistical significance. Calculated peripheral resistance therefore fell in each group and seemed to be the major determinant of the fall in blood pressure occurring with sleep. Despite the tendency to a cardiac output fall in the hypertensive group there was also a tendency for larger reductions in peripheral resistance to be found in the subjects with higher resting pressures. Both the studies of Khatri and Freis and Bristow and colleagues have involved small numbers further subdivided into normotensives and hypertensives and covering individuals of differing ages, sexes, and nationalities and it is hardly surprising that a clear-cut answer to the haemodynamics of sleep has not been forthcoming. It would seem reasonable, however, to suggest that there is a reduction in sympathetic outflow which leads, with variable expression, to a fall in cardiac output and/or a fall in peripheral resistance. This may differ between normotensive and hypertensive groups but until more studies have been performed no clear answer will emerge.

10 Sympathetic Nervous System and Plasma Catecholamines in Sleep

The role of the sympathetic nervous system on blood pressure regulation throughout the night has been investigated recently by Maling et al. (1979) who attempted to correlate blood pressure, heart rate, plasma noradrenaline, and the electro-encephalographic sleep stage in normal subjects. As only five subjects were being studied and as the blood pressure was measured by cuff every 10 min, the results in terms of blood pressure responses are relatively meaningless, but of interest was the finding that plasma noradrenaline levels fell steadily throughout the night (see Fig. 8). However, there is an enormous scatter and as the study was part of a comparison before and after clonidine the samples were not all taken on the same night (i.e. night 3 in some cases and night 5 in others). The similarity of this data, however, to the appearance of the blood pressure record obtained by cuff in Fig. 3 is rather striking. It may be that if the blood pressure had been recorded intra-arterially and the arousal element of the automatic cuff recorder had thus been eliminated, then the plasma noradrenaline would have fallen more abruptly. Clearly these studies must be repeated on subjects where blood pressures are being recorded by means of indwelling arterial cannulae.

11 Blood Gas Changes

It is known that the arterial CO_2 tension rises at night (Birchfield et al. 1958) and this could offer a possible explanation for the fall in blood pressure with sleep by causing peripheral vasodilation and reduction in peripheral vascular resistance. Bristow et al. (1969) also investigated this possibility. They found that Pa CO_2 rose by 4.0+/-1.8 (SD) mm Hg, Pa O_2 fell by 5.4+/-5.1 mm Hg, and pH also fell by 0.014+/-0.012. There was apparently no correlation with blood pressure level, time of night or stage of sleep as determined by the EEG. They do not say whether there was any relationship between the rise in P CO_2 and the fall in blood pressure or calculated peripheral resistance. They do say, however, that the subject with the largest increase in P CO_2 snored the loudest! It would seem unlikely that the arterial P CO_2 is a major determinant of the blood pressure in sleep. The fall takes place so rapidly it is difficult to imagine CO_2 building up to any marked extent in that time (a matter of a few seconds after the alpha rhythm disappears). Hypercapnia would be expected to increase cardiac output and certainly no one has reported that occurring. It is possible that the respiratory centre is less responsive to P CO_2 changes during sleep (Robin et al. 1958). The lack of the various haemodynamic consequences of hypercapnia in sleep would suggest that the vasomotor centre is also less CO_2 responsive at this time.

12 Baroreflex Control and Sleep

Over the past decade or so our laboratory has been interested in baroreflex control of the circulation. It was the observation that nocturnal hypotension is not accompanied by a reflex tachycardia that stimulated the development of a method for assessing baroreflex function in man and indeed the technique was first applied to sleep studies (Smyth et al. 1969). Most physiologists would agree that at least as far as acute changes in blood pressure are concerned the peripheral baroreceptors

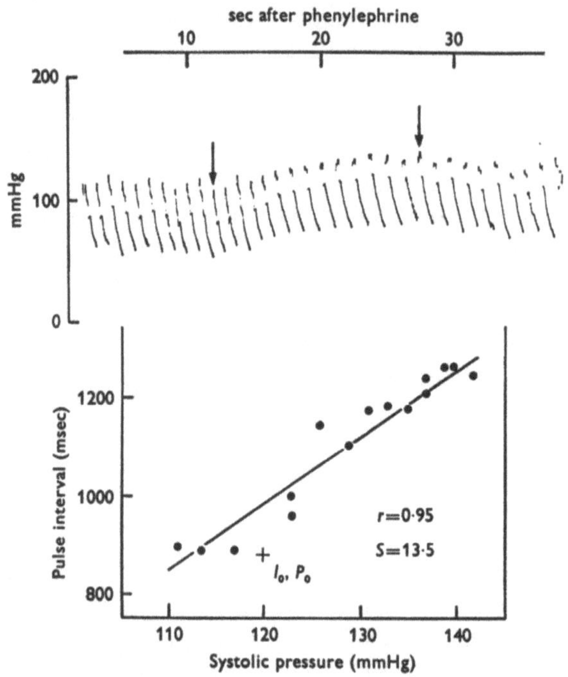

Fig. 9. Determination of baroreflex sensitivity. Above, brachial arterial pressure before, during, and after the rising pressure phase of the response to phenylephrine injection. Below, the relationship between individual pulse intervals I and the systolic pressures P of the same heart beats over the period of the drug-induced pressure rise. The regression line for the filled circles is shown and its slope is the baroreflex sensitivity *S*; *r* is the correlation coefficient. The cross denotes *I* and *P* (I_O, P_O) for the 30-s period before the injection of the drug. (Bristow et al. 1974)

have a dominant role. The question then arises as to why the baroreceptors do not cause an immediate reflex tachycardia when the alpha rhythm disappears and the blood pressure falls briskly.

The drug injection method of assessing baroreflex sensitivity was developed and to understand its use a brief description of the technique is necessary. We quantify the reflex bradycardia which results from an induced rise in pressure provoked by the intravenous injection of a pressor agent. Initially this was angiotensin but latterly has been phenylephrine, a pure alpha receptor agonist. Phenylephrine (50–150 μg) is injected and during the rise in pressure (20–30 mm Hg) the systolic pressure of individual pulses is regressed against the following pulse interval (see Fig. 9). The slope of the regression line obtained in this way gives a measure of baroreflex sensitivity expressed as ms increase in pulse interval/mm Hg rise in blood pressure.

In the initial experiments it seemed that the sensitivity of the baroreflex increased with sleep (Smyth et al. 1969) although in a later study this was not confirmed (Bristow et al. 1969). It was clear, however, that the baroreflex had reset in sleep to operate round the new low blood pressure (Fig. 10) and there was a shift to the left. In both

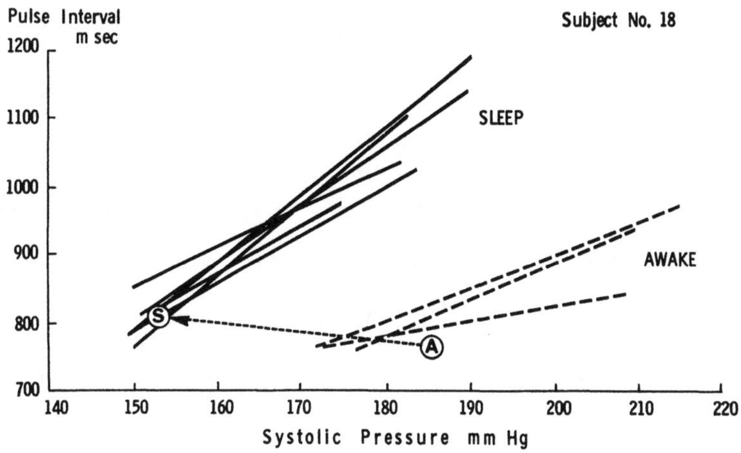

Fig. 10. Baroreflex changes with sleep in one subject. Each line represents one injection of phenylephrine; *A*, average systolic pressure awake; *S*, average systolic pressure during slow wave sleep (Bristow et al. 1969)

studies there was a suggestion that baroreflex sensitivity was increased in REM sleep. The greatest reflex bradycardia for a given injection of phenylephrine occurred in six out of seven cases in the first study (Smyth et al. 1969) and in five out of five in the later one (Bristow et al. 1969) during REM sleep. It is possible, therefore, that baroreflex sensitivity is increased in REM sleep although no consistent changes were seen in sensitivity in the other stages of sleep. Baroreflex sensitivity was not tested in Stage I sleep as this tended to be too brief to allow study.

There is a loss of baroreflex sensitivity with chronic hypertension in awake man (Gribbin et al. 1971). Diminished baroreflex sensitivity was also seen during sleep in hypertension when compared to normotensive controls. However, resetting to the left of the reflex also occurred although the degree of resetting was significantly less in the hypertensive subjects than in normotensive people. The mechanism whereby the reflex is reset in sleep is unknown. Resetting in hypertension is thought to be due to altered mechanical properties of the carotid sinuses and aorta (Aars 1969; Angell-James 1973; Jones 1977; Sleight et al. 1977). In the case of sleep resetting it seems much more likely that there is some alteration within the central nervous system associated with the sleep process itself.

13 K̇-Complexes and Blood Pressure

The EEG during periods of Stage II sleep is distinguished by the frequent occurrence of single fast short waves whose maximum voltage is recorded at the vertex. They are characteristically biphasic, show phase reversal anteroposteriorly and across the midline at the vertex are followed by a slow wave of negativity and are usually attended by bursts of fast waves (sleep spindles) of 12–14 cycles/s, which may either precede

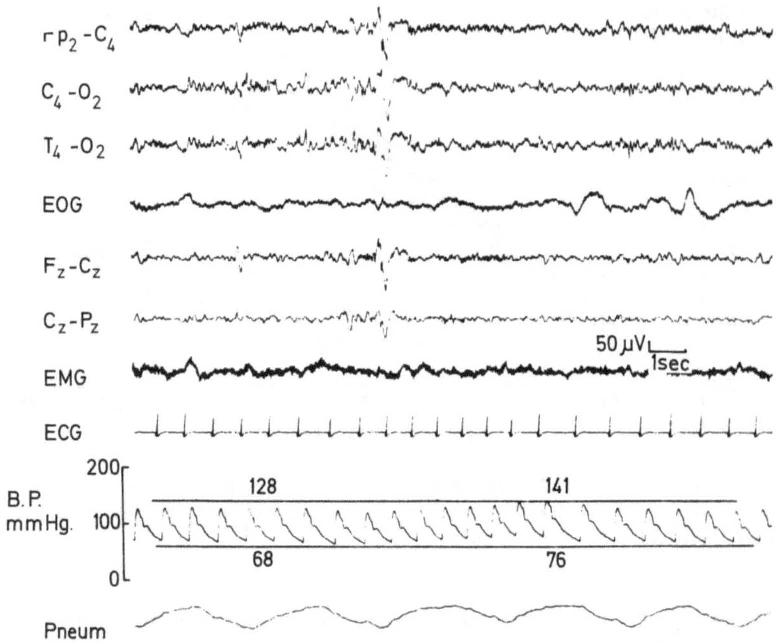

Fig. 11. Circulatory changes following a K-complex. There is a rise in blood pressure accompanied by tachycardia

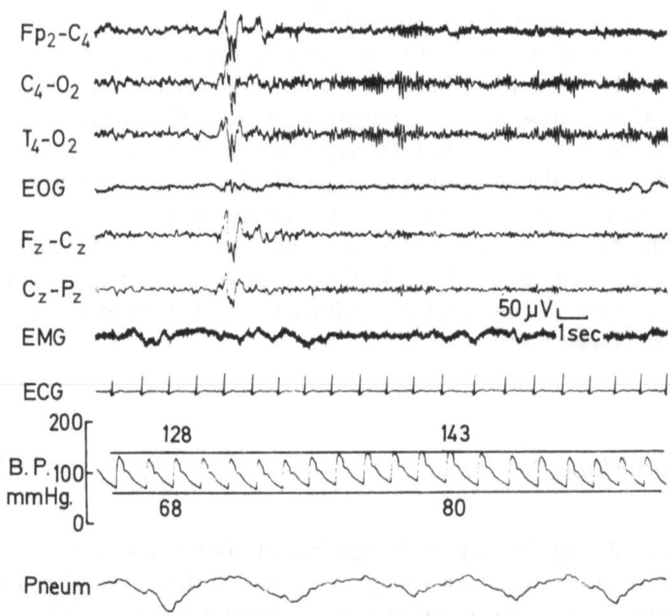

Fig. 12. Following a K-complex there is arousal from Stage II to Stage Ia sleep. There is an increase in blood pressure and heart rate

or follow the short wave. The EEG pattern is called a K-complex (Loomis et al. 1938) and an example can be seen in Fig. 11.

The short wave component may be elicited during light sleep by all forms of sensory stimuli such as light or noise etc. It may lead on to arousal or lightening of sleep (Fig. 12). K-complexes also occur in other stages of sleep but were much more common and more readily apparent in Stage II sleep.

In studies in our laboratory (Smyth 1967) it was found that well over half the number of K-complexes (64.5%) were followed by a rise in blood pressure and brief increase in heart rate (for four or five beats only). The average rise (200 responses in eight subjects) was an increase in systolic pressure of 9.4% and in diastolic of 11.0%. Increases in both rate and depth of respiration also occurred but those were much less frequent and accompanied only about one-third of the pressor responses. The percent rise in pressure was approximately the same whether the K-complex occurred in Stage Ib, II, III, or IV sleep, although, as has been noted, K-complexes are much more frequent in Stage II. However, it did seem that pressure rises are more likely to follow upon K-complexes occurring in Stages Ib and II than upon those occurring in deeper stages of sleep, although the latter were too few to be able to judge this point with certainty. In every case the peak response was reached in less than 10 s, usually within the first six to eight heart beats. The pressures then fell equally as rapidly and control values were reached within another 10 s except where the K-complex led on to arousal with the reappearance of the alpha rhythm on the EEG. In these instances the rises in pressure were usually steeper, higher and more sustained. Examples of typical changes in pressure following K-complexes are shown in Figs. 11 and 12. More sustained elevations of pressure were also observed where several K-complexes occurred in series; it is not unusual for several to follow upon one another at regular intervals, the first having been triggered by a single stimulus.

14 Refractory Period of K-Complex in Man

In man the K-complex has a measurable refractory period and this varies between 1 and 3 s. Roth et al. (1956) showed that a second maximal K-complex might be elicited after 3 s but that the absolute refractory period was about 1 s. Smyth (1967) found that the blood pressure response following a K-complex also had a refractory period and that this was longer than for the K-complex itself. Figure 13 shows two consecutive K-complexes occurring 18 s apart: each is accompanied by a rise in pressure and subsequent return of the pressure to base-line levels. When two such K-complexes occurred within 6 s of each other the second one was not accompanied by a pressor response. The refractory period of the pressure response to a K-complex appears to be about this period of 6 s. The greater length of the refractory period for the blood pressure response following a K-complex compared to the refractory period of the EEG is difficult to explain. It may represent a greater number of synaptic relays on the vasomotor circuit triggered by a K-complex. However, not all K-complexes are followed by pressor responses and it may be that the second of paired K-complexes are in this category.

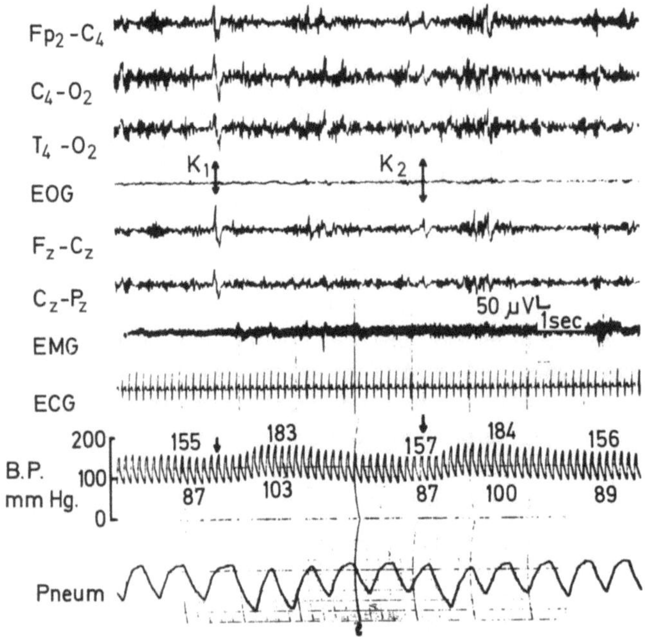

Fig. 13. Identical blood pressure responses follow each of two K-complexes 18 s apart

The observation that heart rate, blood pressure and, less predictably, respiration all increase following a K-complex suggests that there is a brief burst of sympathetic nervous activity. Thus the K-complex could be regarded as an EEG indication of cortical autonomic discharge or simply arousal. As noted earlier Smyth (1967) found that only one-third of the pressor responses following a K-complex were accompanied by an increased depth of respiration. It would seem that the rise in blood pressure cannot then be secondary to respiratory changes. Since both diastolic and systolic pressures are increased with in some cases, diastolic increasing more than systolic, the mechanism of the increase suggests sympathetic vasoconstrictor activity. Support for this comes from Ackner and Pampligione (1955) who found transient diminution of the pulse volume and also finger volume following evoked potentials at the vertex. As long ago as 1897, Howell demonstrated transient vasoconstriction in the hand (using plethysmography) following a noise during uninterrupted sleep. Noise does, of course, readily provoke the appearance of K-complexes on the EEG.

15 Circulatory Changes in Dreaming (REM) Sleep

The few studies that have been made on blood pressure responses to REM sleep have given variable results. There are two good reasons for this. The haemodynamic response to REM sleep appears to be different in animals when compared to man. In human beings the blood pressure response is variable and complex and may change rapidly.

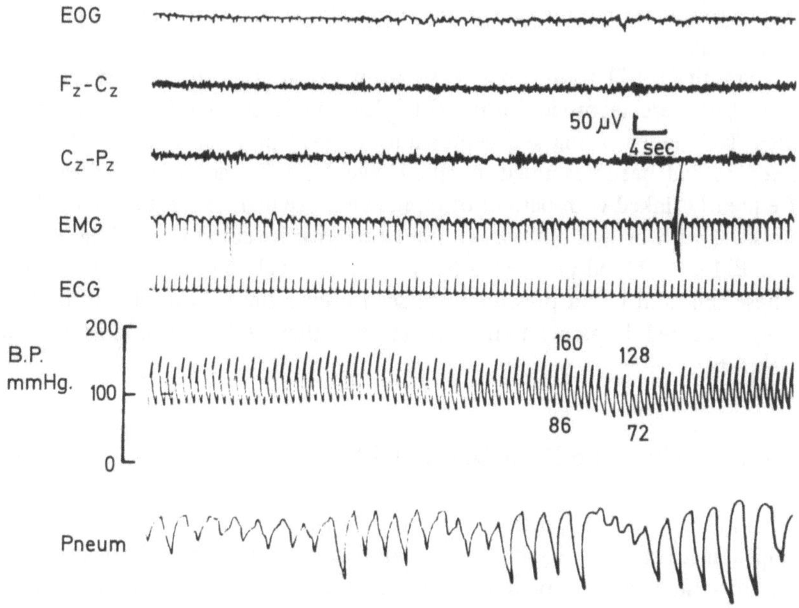

Fig. 14. Shows depressor response in REM sleep. There is brief depression of arterial pressure and respiration associated with a small eye movement (see EOG trace)

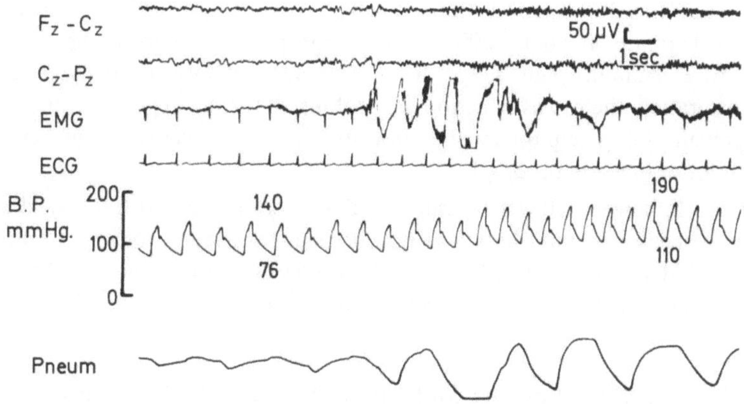

Fig. 15. Shows a pressor response in REM sleep. There is a transient return of the alpha rhythm. Blood pressure rises by 50/30 mm Hg while respiratory amplitude also increases

Hence intra-arterial recordings are mandatory and cuff blood pressure studies of little value.

Smyth (1967) found that in some of his subjects the highest blood pressures of the night were recorded during REM sleep while in others the lowest pressures were seen. No sudden or marked depression of blood pressure attended the onset of REM sleep as had been described in cats. However, after the onset of dreaming sleep, and frequently linked with periods of rapid eye movement, brief, but marked falls in blood pressure with or without concomitant bradycardia could occur (Fig. 14). Nonetheless in REM sleep blood pressure often rose as suddenly and as steeply as it fell (Fig. 15). These spikes in blood pressure were seen intermittently throughout REM periods and they occurred in association with one or other of four events associated with this sleep stage.

16 Blood Pressure Rises During REM Sleep

1. Sudden rises in pressure occurred most commonly in association with brief appearances of alpha rhythm. It was not uncommon for REM periods to be punctuated by frequent short runs of alpha rhythm, usually between bursts of rapid eye movements; the pressure commonly rose for the duration of its appearance (seldom more than 3–4 s).

2. Partial myoclonic jerks (occurring on the limbs or face) were frequently, but not always followed by tachycardia and a rise in pressure.

3. Pressure was seen to rise rapidly in an apparently 'compensatory' overshoot with tachycardia following about 21% of the falls in blood pressure.

4. The most marked rises in pressure associated with REM sleep occurred regularly at the end of each REM period. Periods of dreaming sleep very often ended with a whole body movement and/or a brief arousal with subsequent onset of Stage II sleep; at this transition period blood pressure rose steeply with an increase in heart rate. Furthermore in about 50% of the REM periods observed the subjects performed, during uninterrupted sleep, an involuntary valsalva manoeuvre; following a deep inspiration there was a period of prolonged forced expiration during which arterial pressure fell precipitously and with resumption of respiratory movement a large overshoot of pressure occurred.

Overall, however, lower pressure readings were at least as frequent during REM sleep as higher ones, but there was no consistent tendency for mean pressure to increase in REM periods as Snyder et al. (1964) had reported. These authors had also found the blood pressure during this sleep stage to be extremely variable.

17 Circadian Rhythms and Blood Pressure During Sleep

There seems little doubt that a true circadian rhythm of blood pressure exists. It would be rather surprising if it did not, in view of the many humoral agents known to influence blood pressure that also have a circadian pattern (e.g. renin, corticosteroids, catecholamines etc.) (Akerstedt 1979). Muellar and Brown (1930) collected together nurses' hourly blood pressure charts and found that pressures rose gradually throughout the day to reach a maximum at early evening. The trough of pressures occurred in the middle of the night at about 3 to 4 a.m. Similarly Richardson et al. (1964) found a trough in the day blood pressure at about 5 a.m. with a steady rise thereafter. The subject has received new interest with the introduction of ambulatory monitoring techniques where blood pressures can be monitored continuously on a beat to beat basis in the patient's own home environment. However with such techniques careful note must be made of the patient's activities and due allowance made for these when assessing an individual's blood pressure and not least when trying to determine the presence or otherwise of a circadian rhythm (Watson et al. 1979). As will be immediately apparent, particular note must be taken of times spent asleep and allowance made for this when "pooling" patient data together.

Millar-Craig et al. (1978) found that blood pressure troughed about 3 a.m. and thereafter was followed by a gradual rise in pressure. This rise has come to be termed the "early morning rise" of blood pressure and because many strokes and other vascular disasters have a peak in their incidence about this time, some significance has been attached to this observation. This circadian rhythm has been shown to be present in both normotensive and hypertensive subjects.

18 Intra-arterial Pressure Recordings During Sleep at Home

In our laboratory we have also been performing ambulatory monitoring studies of blood pressure. The technique is as follows: A flexible teflon catheter is inserted into the left brachial artery and connected to a pressure unit by fine tubing. The unit also perfuses the tubing with about, ml h of heparinised saline to keep the cannula patent.

The blood pressure, together with the ECG, are recorded on magnetic tape using an Oxford Medilog four channel recorder. Both recorder and perfusion unit are worn in a comfortable pouch on the chest. The subjects then leave the laboratory and go back to their daily activities returning only for a 15-min period at night for recalibration of the pressure transducer and refilling of the perfusion pump's reservoir. Subjects are instructed to keep an accurate record of their activities with particular attention being paid to the times of retiring and waking. They return after 24 h for removal of the arterial line.

The 24-h record is replayed at 25 times real time into a computer which averages 2 min (or longer) means (and standard deviations) of systolic, mean and diastolic pressures together with heart rate. This can be printed out to give a 24-h blood pressure trace with 720 points (see Figs. 4 and 7). Hourly frequency histograms together with means and standard deviations are also obtained.

We looked to see the effect of arousal, and in this case we meant wakening, in the circadian rhythm of blood pressure (Floras et al. 1978). Fourteen untreated hypertensive subjects were studied and the results shown in Figs. 16 and 17. It can be seen that when the records of all subjects are combined there does seem to be a marked rise in blood pressure starting about 5 a.m. If, however, these same traces are plotted

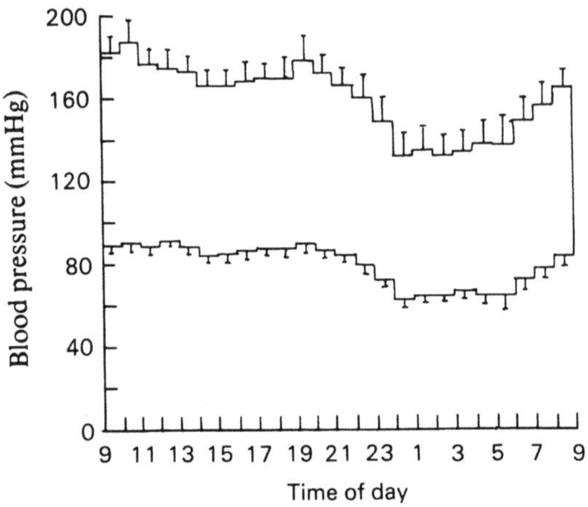

Fig. 16. Mean systolic and diastolic pressures in fourteen untreated hypertensive subjects analysed in hourly intervals over the 24-h period (Floras et al. 1978)

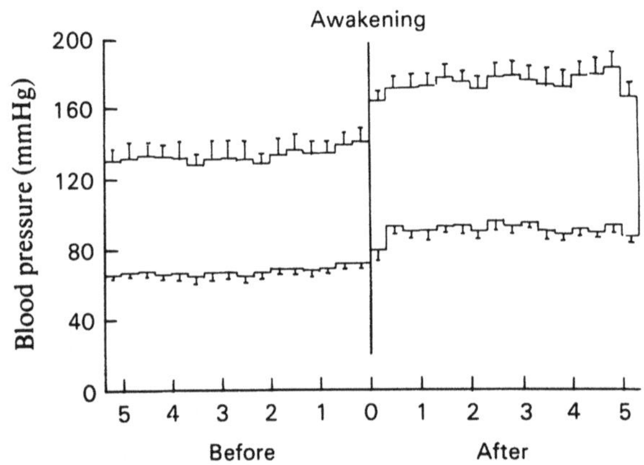

Fig. 17. Data from the same patients as in Fig. 16 analysed with respect to the moment of waking as recorded by the patients, in 20-min intervals for a 5-h period up to the moment of waking and for a five h period subsequent to waking (Floras et al. 1978)

with respect to the time of wakening then this early morning rise is much less marked. There still does seem to be a small rise in pressure, amounting to a few mm Hg over 2 or 3 h prior to arousal (Fig. 17). It would seem from these observations during sleep in home-monitored patients that note must be taken of their activities before describing circadian changes. The ideal experiment would be one where both ambulatory blood pressure and EEG were being recorded simultaneously.

It will be clear from the foregoing discussion that haemodynamic studies during sleep are open to diverse artefactual complications. Blood pressure undoubtedly drops during sleep and rises with arousal. This makes studies utilizing cuff blood pressure measurements of relatively less value than those where direct intra-arterial measurements have been made. Nevertheless much information has been gained using these techniques in the past. Unfortunately few detailed studies correlating the EEG with beat to beat variations in heart rate or blood pressure have been done. Again this, to a large extent, is due to the lack of investigations where direct intra-arterial recordings were performed. Cardiac output measurements are, of course, invasive also and there is clearly a great need for further haemodynamic studies during sleep. Even in hypertension the blood pressure falls with sleep and in many cases this can be quite profound. A greater understanding of the haemodynamic changes which bring this about may lead to greater insight into the "abnormality" of essential hypertension. Although measurement of plasma catecholamines is a relatively poor index of sympathetic activity they are none the less worth performing. Again studies are few and comparisons between normotensive and hypertensive man would be very useful. The newer techniques of ambulatory monitoring where the patient sleeps at home in his own surroundings are helping to give a better understanding of what happens to blood pressure and heart rate during the sleep period. In this regard the "first night effect" is of great importance and any studies except those where the patient has slept at home must be suspect when this effect has not been taken into account. We are currently in our laboratory embarking on more sleep studies with a view to further elucidating the haemodynamic changes which occur during sleep.

References

Aars H (1969) Relationship between aortic diameter and aortic baroreceptor activity in normal and hypertensive rabbits. Acta Physiol Scand 75:406–414

Ackner B, Pampligone S (1955) Combined EEG, plethysmographic respiratory and skin resistance studies during sleep. Electroencephalogr Clin Neurophysiol 7:153

Akerstedt T (1979) Altered sleep/wake patterns and circadian rhythms. Acta Physiol Scand [Suppl] 469:1–48

Angell-James JE (1973) Characteristics of single aortic and right subclavian baroreceptor fiber activity in rabbits with chronic renal hypertension. Circ Res 32:149–161

Bevan AT, Honour AJ, Stott FD (1969) Direct arterial pressure recordings in unrestricted man. Clin Sci 36:329–344

Birchfield RI, Sieker HO, Heyman A (1958) Alterations in blood gases during natural sleep and narcolepsy. A correlation with the electroencephalographic stages of sleep. Neurology (Minneap) 8:107–112

Bristow JD, Honour AJ, Pickering TG, Sleight P (1969) Cardiovascular and respiratory changes during sleep in normal and hypertensive subjects. Cardiovasc Res 3: 476–485

Bristow JD, Brown EB, Cunningham DJC, Howson MG, Lee MJR, Pickering TG, Sleight P (1974) The effects of alveolar P CO_2 and ventilation separately and together on the sensitivity and setting of the baroceptor cardiodepressor reflex in man. J Physiol (Lond) 243:401–425

Brooks H, Carroll JH (1912) A clinical study of the effects of sleep and rest on blood pressure. Arch Intern Med 10:97

Dement W, Kleitman (1957) Cyclic variations in EEG during sleep and their relation to.eye movements, body movements, and dreaming. Electroencephalogr Clin Neurophysiol 9:673–680

Floras JS, Jones JV, Johnston JA, Brooks DE, Hassan MO, Sleight P (1978) Arousal and the circadian rhythm of blood pressure. Clin Sci Mol Med 55:395–397

Gastaut H, Broughton R (1965) A clinical and polygraphic study of episodic phenomena during sleep. In: Wortis J (ed) Recent advances in biological psychiatry, vol 7. Plenum, New York, pp 197–221

Gribbin B, Pickering TG, Sleight P, Peto R (1971) Effect of age and high blood pressure on baroreflex sensitivity in man. Circ Res 29:424–431

Howell WH (1897) A contribution to the physiology of sleep, based upon plethysmographic experiments. J Exp Med 2:313–335

Jones JV (1977) Time-course and extent of carotid sinus baroreceptor threshold resetting in rats with renovascular hypertension. Acta Physiol Scand 99:173–182
Khatri IM, Freis ED (1967a) Haemodynamic changes during sleep. J Appl Physiol 22:867–877
Khatri IM, Freis ED (1967b) Haemodynamics of sleep in hypertensive patients. Clin Res 15:451
Littler WA, Honour AJ (1974) Direct arterial pressure, heart rate, and electrocardiogram in unrestricted patients before and after removal of a phaeochromocytoma. J Med 43:441–449
Littler WA, Honour AJ, Carter R, Sleight P (1975) Sleep and blood pressure. Br Med J 3:346–348
Loomis AL, Harvey EN, Hobart SA (1938) Distribution of disturbance pattern of the human EEG with special reference to sleep. J Neurophysiol 1:413–420
MacWilliam JA (1923) Some applications of physiology to medicine. III. Blood pressure and heart action in sleep and dreams. Br Med J 2:1196–1198
Maling TJB, Dollery CT, Hamilton CA (1979) Clonidine and sympathetic activity during sleep. Clin Sci 57:509–514
Millar-Craig MW, Bishop CN, Raftery EB (1978) Circadian variation of blood pressure. Lancet 1:796–797
Mueller SC, Broom GE (1930) Hourly rhythms in blood pressure in persons with normal and elevated presures. Clin Intern Med 3:1190
Robin ED, Whaley RD, Crump CH, Travis DM (1958) Alveolar gas tensions, pulmonary ventilation, and blood pH during physiologic sleep in normal subjects. J Clin Invest 37:981–989
Richardson DW, Honour AJ, Fenton GW, Stott FH, Pickering GW (1964) Variations in arterial pressure throughout the day and night. Clin Sci 26:445
Richardson DW, Honour AJ, Goodman AC (1965) Central nervous control of blood pressure in man (Preliminary report). Med Coll Virginia Q 1:24–27
Roth M, Shaw J, Green J (1956) The form voltage distribution and physiological significance of the K-complex. Electroencephalogr Clin Neurophysiol 8:385
Shaw DB, Knapp MS, Davis DH (1963) Variations of blood pressure in hypertensives during sleep. Lancet 1:797–798
Sleight P, Robinson JL, Brooks DE, Rees PM (1977) Characteristics of single carotid sinus baroreceptor fibers and whole nerve activity in the normotensive and the renal hypertensive dog. Circ Res 41:750–758
Smirk FH (1957) High arterial pressure. Blackwell, Oxford, p 65
Smyth H (1967) Sleep, dreams, and blood pressure. PhD thesis, Oxford University
Smyth HS, Sleight P, Pickering GW (1969) The relfex regulation of arterial pressure during sleep in man; a quantative method of assessing baroreflex sensitivity. Circ Res 24:109–121
Snyder F, Hobson JA, Morrison DF, Goldfrank F (1964) Changes in respiration, heart rate, and systolic pressure in human sleep. J Appl Physiol 19:417
Watson RDS, Stallard TJ, Littler WA (1979) Factors determining the variability of arterial pressure in hypertension. Clin Sci 57:283–285

Subject Index

Current Topics in Neuroendocrinology

Volume 2:

Adrenal Actions on Brain

1982. 25 figures. Approx. 180 pages
ISBN 3-540-11126-3

Contents: Glucocorticoids and Hippocampus: Receptors in Search of a Function. - Glucocorticoids and the Developing Nervous System. - Feedback Actions of Adrenal Steroid Hormones. Electrophysiologic and Clinical Aspects of Glucocorticoids on Certain Neural Systems. - Adrenal Steroids and Behavioral Adaptation: Relationship to Brain Corticoid Receptors. - Subject Index.

The central nervous system and the endocrine system comprise quickly expanding fields at the forefront of medical and biological research. Neuroendocrinology is pursued so intensely that it has become impossible to keep up to date on all current research activity.
Current Topics in Neuroendocrinology helps overcome this problem by treating not only those topics which are exciting because of their topicality, but also those which are reliable, where enough work has been done to treat them from several viewpoints. This volume covers the relations between the adrenal glands and the brain, with emphasis on the regulatory effects of the glucocorticoids. The articles it contains review recent work or hormone receptors and their developmental, endocrinological, electrophysiological and behavioral effects on the central nervous system.

Springer-Verlag
Berlin
Heidelberg
New York

J.T. Enright

The Timing of Sleep and Wakefulness

On the Substructure and Dynamics of the Circadian Pacemakers Underlying the Wake-Sleep Cycle
With a Foreword by E. Flory
and an Appendix by J. Thorson
1980. 103 figures, 2 tables. XVIII, 263 pages
(Studies of Brain Function, Volume 3)
ISBN 3-540-09667-1

Contents: Introduction. - A Description of Activity-rhythm Recordings and Their Implications. - The Pacemaker and Its Precision. - A Class of Models for Mutual Entrainment of an Ensemble of Neurons. - A "Type Model" and its Behavior: Partial and Loose-Knit Mutual Entrainment. - Precision of Model Pacemakers. - Influences of Constant Light Intensity. - A Brief Detour: Further Thoughts About the Discriminator of the Models. - General Features of Entrainment: The Type Model. - Responses to Single Light Pulses. - Part I: Nocturnal Rodents. - Responses to Single Light Pulses. Part II: Diurnal Birds. - Plasticity in Pacemaker Period: A Dynamic Memory. - Predictions from Coupled Stochastic Systems. - Further Predictions: A Modest Success and Two Problem Cases. - Morphology of the Models: Where is the Pacemaker? - A Reprise and Synopsis: On the Advantages of Apparent Redundancy. - References. - Author Index. - Subject Index.

The timing of the wake-sleep cycle of higher vertebrates is controlled by a pacemaker which is assumed to be located in the brain. This book, intended for a broad audience at the graduate-student level, describes the computer-assisted exploration of certain hypotheses about that pacemaker. Interest is focussed upon how single cells - possibly neurons with an independent capacity for long-period oscillations - might be interconnected and might interact to produce the overt rhythm. A simple sort of neuronal network is proposed, the elements of which have dynamics formally similar to those of tonically-active neurons in the high-frequency domain. Stochastic (i.e. random) processes, as well as threshold effects, play prominent roles in the system. Such an ensemble can account for the remarkable temporal precision often seen in the experimental data: many unreliable cells which entrain each other can constitute a very precise "clock". In addition, models of this sort can account naturally and in detail for a large body of data on other aspects of the circadian rhythms of birds and rodents, ranging from various influences of light intensity to long-term plasticity in rhythm properties, and can leed to many shorts of predictions.

Springer-Verlag
Berlin
Heidelberg
New York

MIX
Papier aus verantwortungsvollen Quellen
Paper from responsible sources
FSC® C105338

If you have any concerns about our products,
you can contact us on
ProductSafety@springernature.com

In case Publisher is established outside the EU,
the EU authorized representative is:
Springer Nature Customer Service Center GmbH
Europaplatz 3, 69115 Heidelberg, Germany

Printed by Libri Plureos GmbH
in Hamburg, Germany